Electronic Devices, Circuits, and Applications

Christopher Siu

Electronic Devices, Circuits, and Applications

 Springer

Christopher Siu
British Columbia Institute of Technology
Burnaby, BC, Canada

ISBN 978-3-030-80540-1   ISBN 978-3-030-80538-8   (eBook)
https://doi.org/10.1007/978-3-030-80538-8

This Springer imprint is published by the registered company Springer Nature Switzerland AG
The registered company address is: Gewerbestrasse 11, 6330 Cham, Switzerland

To Maki, Nick, and Rina

Preface

What is the purpose of yet another electronic textbook? The short answer is the usage of a different teaching paradigm and modern simulation tools. This book is based on lectures for a first course in electronics that I teach at the British Columbia Institute of Technology. The course is taken by first year students who may continue to complete a 2-year diploma or a 4-year degree. The challenge is to present transistors in a way that is useful for both technologists and engineering students.

Through my own experience as a student, I remember the study of electronics and analog circuits as being far from simple. University-level textbooks typically start with the device physics of diodes and transistors, before proceeding to equations of their characteristics. For an introductory course, while the terminologies of semiconductor physics are useful to know, a detailed explanation at this point only serves to confuse the beginner. What I felt was needed is a book that focuses on having simple models for diodes and transistors, and that teaches a beginner how to use them in actual circuits. This is the impetus behind this work.

The book uses a technique called the *spiral* approach, in that the same topic is presented multiple times, but at increasing levels of detail. Each chapter is kept short to divide information into understandable segments. For example, we start with a simple 0.7V model for a silicon diode and work with that for an entire chapter. A subsequent chapter then explores the diode in more detail, such as Shockley's equation and its device physics. This approach not only helps to build and reinforce concepts, but also allows the learner to pick and choose topics at different levels.

In addition, the availability of free circuit simulation tools has greatly increased in the last two decades. One tool that I have adopted for class use, in the midst of the pandemic, is a free tool called Tinkercad® (www.tinkercad.com) by Autodesk®. This tool uses a graphical interface with virtual breadboards and test equipment, allowing students to conduct virtual labs anywhere with an internet connection.

I hope that this work is accessible to a wide audience who is interested in learning about electronics.

Vancouver, Canada   Christopher Siu
March 2021

Contents

Interpreting I-V Curves

In working with semiconductor devices, we will encounter I-V curves, which are graphical representation of the current-vs-voltage (I-V) behavior of a device. In fact, there are commercial equipment that are used to measure I-V curves of diodes and transistors, which beg the question: how do we read an I-V curve? We will start by looking at the I-V curves of a few simple circuit elements.

1 Ideal Voltage Source

An independent voltage source would ideally maintain a fixed voltage across its terminals regardless of the current. For example, an ideal +5 V source will force node n1 to be 5 V higher than node n2, regardless of whether +10 A, 0 A, or −5 A is drawn from this voltage source. Plotting these points on a graph of current vs voltage, we would obtain a vertical line for an ideal voltage source (Fig. 1).

2 Ideal Current Source

An independent current source would ideally maintain a fixed current regardless of the voltage across it. For example, an ideal 2A current source will force 2A to flow from n1 to n2, regardless of the voltage V across these two nodes. Plotting points with y-coordinates fixed at 2A yield a horizontal line on the I-V graph (Fig. 2).

© Springer Nature Switzerland AG 2022
C. Siu, *Electronic Devices, Circuits, and Applications*,
https://doi.org/10.1007/978-3-030-80538-8_1

Fig. 1 I-V curve of an
ideal +5 V voltage source

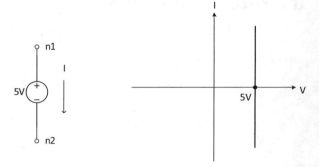

Fig. 2 I-V curve of an
ideal 2A current source

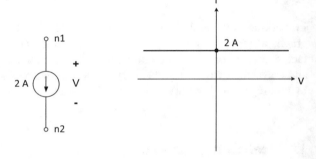

3   Resistors

An ideal resistor obeys Ohm's law, which can be written as an equation where current is a linear function of voltage:

$$I = \frac{1}{R} \cdot V$$

Hence, the I-V curves of resistors are straight lines, where the slope of the line is the inverse of resistance. For a large resistance value, the slope will be small, meaning that the I-V curve of a large resistor approaches a horizontal line (Fig. 3).

Since the slope of an I-V curve is equal to the conductance $G = 1/R$ of the device, we can now relate this to the resistance of an ideal voltage source and ideal current source (Table 1):

Table 1 Ideal voltage and current sources comparison

	Slope of I-V curve	Resistance (Ω)
Ideal voltage source	Infinite	0
Ideal current source	0	Infinite

Fig. 3 I-V curve of a resistor

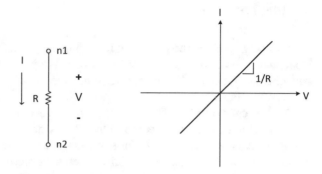

Resistances of an ideal voltage source and ideal current source will be used repeatedly in later analysis, and so it is worth understanding and memorizing this idea before moving on.

Exercise Sketch the I-V curve of an ideal −1.5 A current source

Answer A horizontal line intersecting the y-axis at −1.5 A

Exercise What is the resistance represented by the I-V curve below?

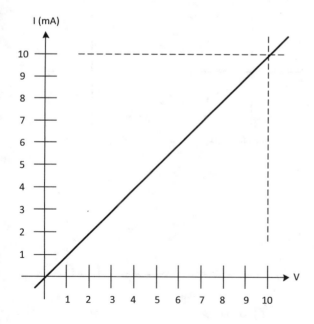

Answer 1 kΩ

4   Ideal Switch

We will now combine the preceding ideas into an I-V curve of an ideal switch. An ideal switch has two states: opened or closed. When the switch is open, the resistance between its terminals is infinite. When the switch is closed, the resistance is ideally zero (Fig. 4).

The I-V curve on the left can be used to represent an opened switch. Since the line is horizontal, it is equivalent to an infinite resistance, and most importantly, it is telling us that the current is zero for any voltage across the switch.

Similarly, the I-V curve on the right represents a closed switch. The vertical line is equivalent to a zero resistance, and the voltage across the switch is zero for any current flowing through the switch.

In a mechanical switch, we have to manually flip it between the opened and closed positions. Now imagine that we have a special switch that can be controlled by an electrical signal. For this special switch, the voltage across it determines whether it is opened or closed (Fig. 5).

Fig. 4 I-V curve of an
ideal switch (**a**) opened (**b**)
closed

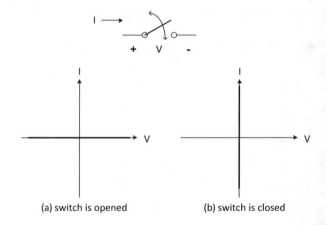

(a) switch is opened   (b) switch is closed

Fig. 5 I-V curve of an
electronic switch

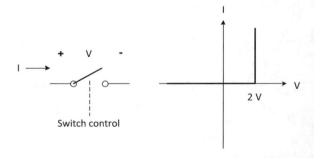

In the example above, the switch is opened when the voltage V across it is less than 2 V. Once the voltage V reaches +2 V, the switch automatically closes. However, the one thing different from an ideal switch is that when this special switch is closed, there is a 2 V drop; as seen from the I-V curve, this 2 V drop is fixed regardless of the current. Another thing that is implied by this I-V curve is that current only flows in one direction through the switch, from left to right; given the +2 V drop across this closed switch, current must flow from left to right by convention.

The idea of using an electrical signal to control a switch is a key application of diodes and transistors.

5   DC Resistance vs AC Resistance

We have seen how the slope of an I-V curve is equivalent to the conductance of the device. To make sense of the transistor models in later chapters, we need to understand the idea of DC resistance and AC resistance. Both DC resistance and AC resistance are based on Ohm's law, and this section will outline how these quantities are different from each other.

For a straight line, the slope is constant at every point, and hence it is simple to find the resistance of device. For example, the I-V curve of a 10 Ω resistor is shown below (Fig. 6).

Fig. 6 I-V curve of a 10 Ω resistor

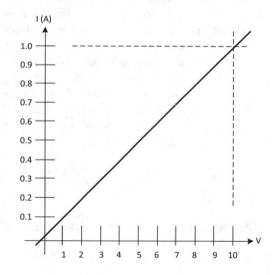

Definition of DC Resistance

$$R_{DC} = \frac{V}{I}$$

To find the DC resistance, we take one point on the I-V curve and perform the calculation. For example, in the graph above, the current is 0.1 A for a voltage of 1 V. Hence, $R_{DC} = 1\ \text{V} / 0.1\ \text{A} = 10\ \Omega$.

Definition of AC Resistance

We had observed that the slope of an I-V curve is equivalent to resistance, which we will now formally define as the AC resistance.

$$r_{ac} = \frac{\Delta v}{\Delta i}$$

To find the AC resistance, we are looking for the slope or tangent to a point on the I-V curve. To find the slope of a straight line is easy enough; given a change Δv in the independent variable, we find the change Δi in the dependent variable. Referring to the I-V curve above, we have points such as (0.1 A, 1 V) and (0.9 A, 9 V). The calculation of AC resistance yields $(9\ \text{V} - 1\ \text{V})/(0.9\ \text{A} - 0.1\ \text{A}) = 10\ \Omega$.

For a resistor, the DC resistance and AC resistance are equal to each other. We can make the same conclusion by seeing that a resistor's I-V curve is a straight line.

What happens if the I-V curve is not a straight line? (Fig. 7)

In general, a nonlinear I-V curve will yield different DC resistance and AC resistance values. In the example below at +1 V, the current is 0.1 A, and $R_{DC} = 1\ \text{V}/0.1\ \text{A} = 10\ \Omega$. The AC resistance, on the other hand, requires that we find the slope of this curve at +1 V. If we have an equation that described the voltage-to-current relationship, we can take a derivative to find the slope at that point. In this case, no equation is available, but we do have the I-V curve and data as shown below. The slope can be estimated by using two points near each other.

$$\text{Slope} = \frac{\Delta i}{\Delta v} = \frac{0.121\,\text{A} - 0.1\,\text{A}}{1.1\,\text{V} - 1\,\text{V}} = \frac{0.021\,\text{A}}{0.1\,\text{V}}$$

AC resistance is the inverse of the slope, hence

$$r_{ac} = \frac{\Delta v}{\Delta i} = \frac{0.1V}{0.021A} = 4.76\,\Omega$$

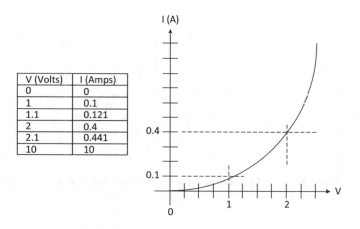

V (Volts)	I (Amps)
0	0
1	0.1
1.1	0.121
2	0.4
2.1	0.441
10	10

Fig. 7 I-V curve of a non-linear device

Exercise Find the DC resistance and AC resistance at +2 V for the nonlinear I-V curve shown previously

Answer $R_{DC} = 5\ \Omega$, $r_{ac} = 2.44\ \Omega$

Semiconductor devices have nonlinear I-V characteristics, and this is a key reason why we start with the study of I-V curves. From the previous example, we can make some important observations:

- The DC resistance and AC resistance values are generally different from each other for a nonlinear device
- Since the slope of the I-V curve is not constant, the AC resistance also varies depending on where we are on the curve

Though it may seem like an odd idea, AC resistance is trying to quantify how a device responds to *changes* in voltage. This is a precursor to the concept of impedance, which relates changes in current to the changes in voltage over time. We will see how AC resistance is used later on, in the analysis of transistor amplifiers.

6 Problems

1. What is the AC resistance of an ideal voltage source?
2. What is the AC resistance of an ideal current source?
3. The electronic switch described in this chapter is shown again below:

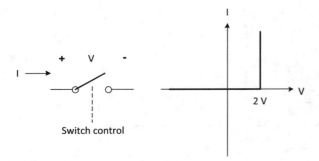

Find the voltage and current through the switch in the following circuits:

(a)

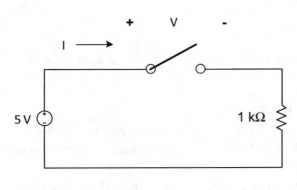

(b)

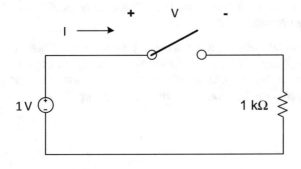

(c)

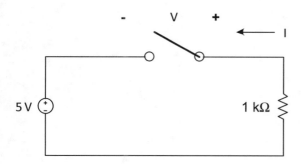

4. You are given a device whose voltage-to-current relationship is given by the following equation:

$$I_a = 1 \ mA \cdot V_a^2$$

(a) Find the DC and AC resistances at $V_a = 1$ V
(b) Find the DC and AC resistances at $V_a = 2$ V

5. You are a given a device whose voltage-to-current relationship is given by the following equation, where k = 5:

$$I_b = 1 \ \mu A \cdot e^{k \cdot V_b}$$

(a) Find the DC and AC resistances at $V_b = 1$ V
(b) Find the DC and AC resistances at $V_b = 2$ V

Introduction to the Diode

Solid-state devices, or devices that control electricity through solid semiconductor materials, are relatively recent inventions that were made possible by the development of quantum mechanics. The detailed treatment of semiconductor device physics is a difficult topic requiring advanced mathematics, which is not a good place to start. Fortunately, for circuit designers using solid-state devices, a simple behavioral model is a good starting point, and that is the approach that we will take in the introduction to various devices.

The first semiconductor device we will study is the silicon diode. A diode is a device that only allows current to flow in one direction, and we will see how this property is useful in a number of applications. To begin, the symbol for a diode and the names of its two terminals are shown below (Fig. 1).

1   A Simple Diode Model

Recall from the I-V curve chapter that we created the plot of an ideal switch, where the switch would open or close depending on the voltage across it. The simple model of a silicon diode is the same except for the switch control voltage. If the voltage across the diode V_D is less than 0.7 V, the diode is off and does not conduct. Once V_D reaches 0.7 V, the diode turns on, and current will flow in the direction indicated (Fig. 2).

Before moving on to some examples, we need to be familiar with these terminologies (Fig. 3):

If the voltage V_D is positive, the diode is said to be in forward bias. Note that if V_D is positive but less than 0.7 V, then according to our simple model the diode is still off and conducts no current. Once V_D is equal to 0.7 V, the diode is on and conducts current in the direction shown. Regardless of the magnitude of the current, the diode voltage V_D is fixed at 0.7 V.

© Springer Nature Switzerland AG 2022   11
C. Siu, *Electronic Devices, Circuits, and Applications*,
https://doi.org/10.1007/978-3-030-80538-8_2

Fig. 1 Diode symbol and terminal names

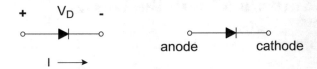

Fig. 2 Diode simple model I-V curve

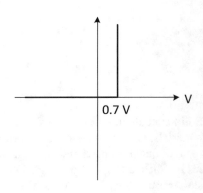

Fig. 3 Diode forward bias & reverse bias

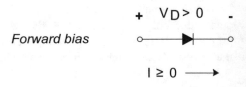

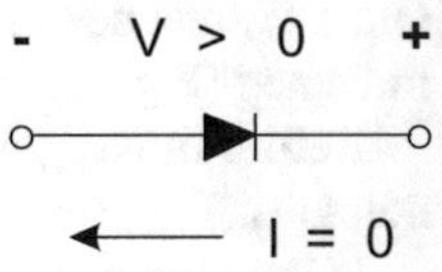

Fig. 4 Alternate representation of diode reverse bias

If the voltage V_D is negative, the diode is in reverse bias and conducts no current. An equivalent way of drawing this is to reverse the voltage polarity; this is a common way of annotating circuits so that we can clearly see the voltage is creating a reverse bias (Fig. 4).

Exercise Find the quantities V and I as indicated in circuit below

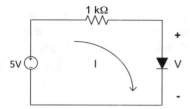

Answer The first step in solving DC circuits with diodes is to ask the question "Is the diode on or off?." In this circuit, we can see that 5 V will be dropped across the resistor and the diode. Hence, there is enough voltage to turn the diode on, leaving 5 V − 0.7 V = 4.3 V across the resistor. The current $I = 4.3$ V/1 kΩ = 4.3 mA. The voltage $V = V_D = 0.7$ V.

Exercise Find the quantities V and I as indicated in the circuit below.

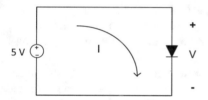

Answer By KVL, there is supposed to be 5 V across the diode in forward bias. According to the simple diode model, this is not possible, so this circuit does not work. In real life, if you built this in this lab, the diode will likely explode when you turn on the 5 V power supply.

Exercise Find the quantities V and I as indicated in circuit below.

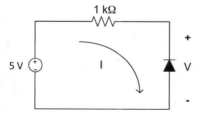

Answer Is the diode on or off? We can see that the diode is reverse biased, and hence it is off; $I = 0$ and $V = 5$ V. Note that the voltage polarity V is opposite to the definition of V_D or $V_D = -V$. Hence, $V_D = -5$ V.

2   Diode Applications in DC Power Supplies

One of the key applications of diodes is AC to DC conversion, otherwise known as rectification. All electronics that take power from a wall socket (AC) has a rectifier to convert the line voltage to a DC voltage; the DC voltage is then used to operate the integrated circuits and other components in the system.

Traditional DC Power Supplies

Not long ago, a DC power supply used a step-down transformer to convert the AC line voltage to a lower AC voltage, followed by a rectifier and other components to create a regulated DC output. The schematic symbol for a transformer is shown below (Fig. 5).

The transformer can be viewed as an AC to AC converter: it accepts an AC voltage in the primary and outputs an AC voltage with a different amplitude on the secondary. The frequency of the input and output is the same. In North America, the AC line voltage from a wall socket is 120 V_{RMS} at 60 Hz. The line voltage is applied to the primary, and in a step-down transformer, the secondary voltage is lower than the line voltage (e.g., 20 V_{RMS}). Note that the average values of the primary voltage and the secondary voltage are zero.

Next, a rectifier is connected to the transformer secondary. This circuit takes secondary voltage (60 Hz sine wave) and performs an absolute value operation on it. As a result, the rectifier output has a non-zero average, or a positive DC value. A rectifier can thus be viewed as an AC to DC converter (Fig. 6).

The rectifier output, while it has a non-zero DC average, has a large voltage ripple versus time. These ripples must be reduced before the output can be used as a power supply for integrated circuits. As shown in the block diagram below, a capacitor is used to reduce ripples; we will see how to size this capacitor later in the chapter. Finally, a voltage regulator converts the capacitor voltage to some desired level (e.g., 5 V), with virtually no ripple at the final output (Fig. 7).

Fig. 5 Transformer symbol

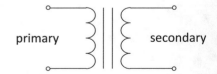

primary   secondary

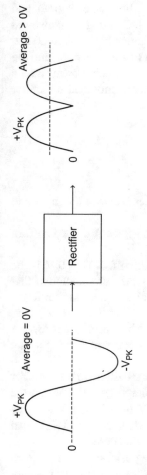

Fig. 6 Operation of a full wave rectifier

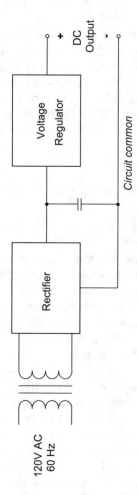

Fig. 7 Traditional DC power supply: block diagram

Modern Offline Power Supplies

Modern power supplies such as your smartphone charger is based on switching topologies that yield a much higher efficiency, while being small and lightweight. In traditional power supplies, a transformer is connected directly to the AC line, and since this transformer operates at the low frequency of 60 Hz, it is rather bulky and heavy (the reason for this is based on the physics of electromagnetism, which will not be explained here).

In a modern supply, a rectifier is connected to the AC line. The rectifier and capacitor create a high DC voltage with some ripple. This DC voltage is converted into a square wave via a switch that toggles at a frequency much higher than 60 Hz, such as 100 kHz. The high voltage square wave is the input to a transformer, and the key point is that due to the high frequency, the transformer is much *smaller* than a 60 Hz transformer. The transformer secondary voltage is then smoothed by a filter to produce the desired DC output (Fig. 8).

3   Half Wave Rectifier

The simplest rectifier involves one diode connected in series between the input and the output (Fig. 9).

To understand how this circuit works, let us think about V_{out} if the diode is off. In this case, the diode is like an open switch, which means $V_{out} = 0$ V.

Next, we examine what values of V_{in} would turn the diode off. If $V_{in} = 0$ V, is the diode off? Starting with the assumption that the diode is off, we concluded that $V_{out} = 0$ V. If $V_{in} = 0$ V, then the diode voltage $V_D = 0$ V, and thus our assumption is correct. If $V_{in} = 0$ V, the diode is off.

Following the same logic, we can see that if V_{in} goes negative, the diode is reverse biased and remains off; hence for the entire negative half cycle of the input, the diode is off and $V_{out} = 0$ V (Fig. 10).

When V_{in} is positive, then it must be 0.7 V or greater for the diode to turn on. For example, if $V_{in} = +0.5$ V and $V_{out} = 0$ V, then the diode voltage $V_D = 0.5$ V and the diode is off. On the other hand, if $V_{in} = +2$ V and $V_{out} = 0$ V, then $V_D = 2$ V and this is invalid using the simple model; for a forward biased diode that is on, V_D is fixed at 0.7 V. Hence, for $V_{in} = +2$ V, $V_{out} = V_{in} - 0.7$ V $= 1.3$ V in this case.

Based on the above analysis, we conclude that only the positive half of V_{in} goes to the output, with a 0.7 V drop as shown below. Half of the input sine wave is transferred to the output, hence the name half wave rectifier.

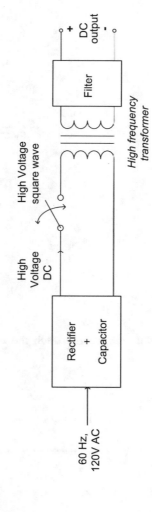

Fig. 8 Off-line DC power supply: block diagram

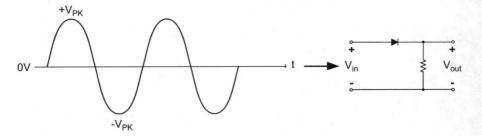

Fig. 9 Half wave rectifier circuit

Fig. 10 Operation of a
half wave rectifier

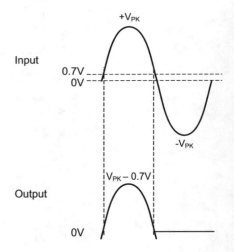

Exercise For a 10 V peak input sine wave, sketch the output of the half wave rectifier.

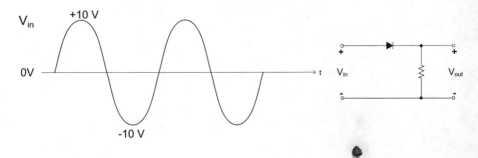

Answer The circuit allows only the positive half of the sine wave through to the output with a 0.7 V drop. Hence, the output peak is 10 V − 0.7 V = 9.3 V

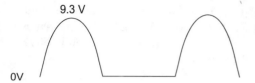

9.3 V

0V

Exercise For a 10 V peak input sine wave, sketch the output of the circuit below

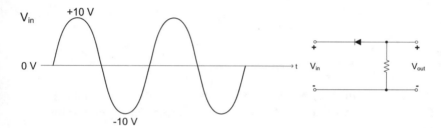

V_{in} +10 V

0 V   →t V_{in} V_{out}

-10 V

Answer Compared to the original circuit, the direction of the diode has been reversed. By going through the analysis procedure described previously, we can see that the negative half of the input will go to the output with a 0.7 V drop

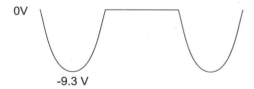

0V

-9.3 V

Exercise For a 50 V peak input sine wave, sketch the voltage across the diode versus time.

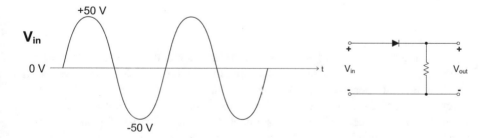

+50 V

V_{in}

0 V   →t V_{in} V_{out}

-50 V

Answer Note that the question is asking not for the output, but the diode voltage V_D. We know that if $V_{in} \geq 0.7$ V, then the diode is on; for most of the positive half cycle, $V_D = 0.7$ V. On the other hand, when V_{in} is negative, the diode is off. For example, for $V_{in} = -50$ V, the diode voltage $V_D = -50$ V.

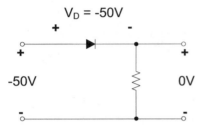

Hence, when the diode is off, V_D will follow V_{in}, and in this example the diode is off for V_{in} between -50 V to 0.7 V. The overall waveform for the diode voltage is shown below.

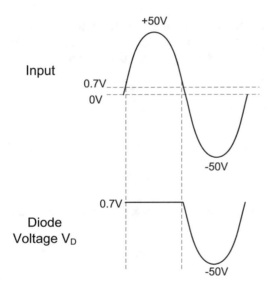

From the previous example, we can see that the maximum reverse bias corresponds to the negative peak of V_{in}. According to the simple model, a reverse-biased diode is like an open switch. However, we can guess that there is a limit on this reverse voltage; if the diode voltage is -1000 V, the diode may not behave like an open switch. In a real diode, there is a datasheet parameter called peak repetitive reverse voltage, also known as the breakdown voltage, which indicates the limit on the diode reverse bias. Once the breakdown voltage is exceeded, the diode will conduct in the reverse direction. This can be incorporated into the simple diode model as shown below (Fig. 11).

Fig. 11 Diode simple model with reverse breakdown

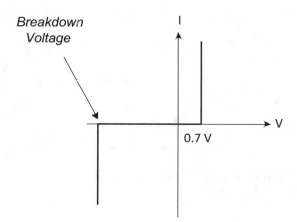

4   Full Wave Rectifiers

The half wave rectifier, while simple, retains only half of the input; the negative half cycle of the input is blocked and discarded. If we wish to use both halves of the input sine wave, we will need to use a full wave rectifier circuit.

Full Wave Bridge Rectifier

The first full wave rectifier we will study is called a bridge rectifier. Below are two equivalent schematics of this circuit (Fig. 12).

To understand the operation of the bridge rectifier, we will break the analysis into two steps: we first examine which diodes are on when V_{in} is positive and then which diodes are on when V_{in} is negative. To aid in the analysis, we recall that current can only flow through the diode in one direction, in the same direction that diode symbol triangle is pointing at (Fig. 13).

Positive Half Cycle of Input   Over the interval T1 when V_{in} is positive, current flows into the top terminal and out of the bottom terminal by convention. Looking at the top terminal, current into that node can flow through D1 but not D2. Next, the current out of D1 cannot go into D4; hence it must flow through the load resistor and generate V_{out}. Finally, the current through the load resistor must return to the bottom terminal, and D3 must be on for that to happen. Note that the current travels through two diodes, D1 and D3, when V_{in} is sufficiently positive; as a result, the output is reduced by two diode drops (2×0.7 V $= 1.4$ V) compared to the input (Fig. 14).

Negative Half Cycle of Input   Over the interval T2 when V_{in} is negative, current flows into the bottom terminal and out of the top terminal by convention. Using the same ideas as before, D4 and D2 are now on; note that the current flows through the load resistor in the same direction even though V_{in} is negative. Hence, the output is the same as what the positive half cycle produced, as shown in the diagram below (Fig. 15).

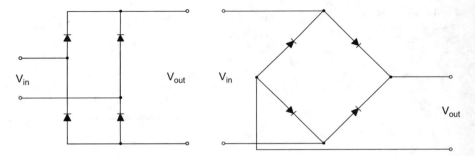

Fig. 12 Full wave bridge rectifier circuit

Fig. 13 Diode
current flow

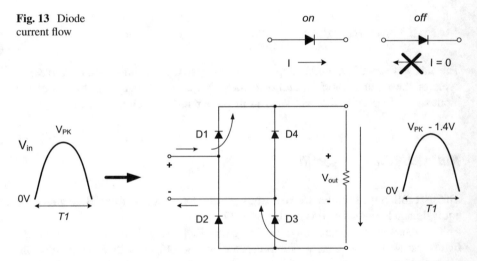

Fig. 14 Bridge rectifier analysis: positive half cycle

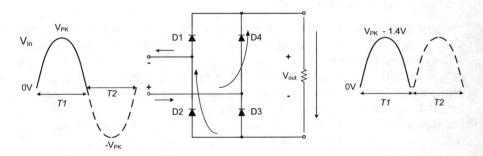

Fig. 15 Bridge rectifier analysis: negative half cycle

A key thing to note is that if the input is 60 Hz, the rectified output is **120 Hz**. The interval T1 + T2 is equal to one period of the 60 Hz sine wave, whereas T1 (or T2) is one period of the rectifier output.

Exercise If the input to a bridge rectifier is connected to the line voltage in North America (120 V_{RMS}), sketch the rectifier output and label the peak voltage on the output.

Answer First we need to sketch the rectifier input, which is a 60 Hz sine wave at 120 V_{RMS}. Conversion from RMS to peak requires multiplying by $\sqrt{2}$; hence the input has a peak of 120 × $\sqrt{2}$ = 170 V. The input waveform and the circuit are shown below.

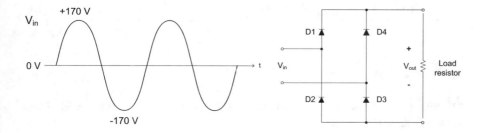

Next, we recognize that the bridge rectifier produces an output that is the absolute value of input, minus two diode drops of 2 × 0.7 V = 1.4 V. The output of the rectifier is shown below:

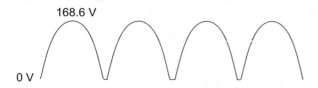

Full Wave Center-Tapped Rectifier

The bridge rectifier requires four diodes, and during operation, there are two diode drops going from the input to the output. However, we can use a different circuit topology if a center-tapped transformer is available. Physically, a center tap is created by making a connection to the middle of the secondary winding. Electrically, by using the center tap as circuit common, we can build a full wave rectifier using only two diodes (Fig. 16).

Center-tapped transformers are specified with CT appended to their nominal secondary voltages, such as 20VCT. One important thing to note is that this specification refers to the RMS voltage across the *entire* secondary. If we measure the voltage

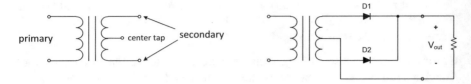

Fig. 16 Center tapped transformer and rectifier circuit

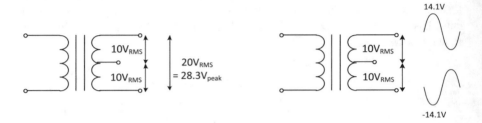

Fig. 17 Center tapped transformer secondary voltages

between one of the secondary terminals and the center tap, then we would obtain half the voltage (Fig. 17).

Note that when the upper secondary terminal is positive with respect to (wrt) the center tap, the lower secondary terminal is negative. This means that D1 can turn on, but D2 is off; hence the transformer secondary current flows through D1 to the load.

When the upper secondary terminal is negative wrt to the center tap, the lower terminal is positive. This means that D1 is off, D2 is on, and the secondary current flows through D2 to the load.

Example Given a 20VCT transformer, sketch the output of a full wave center tapped rectifier.

Answer We had looked at the 20VCT transformer in the main text. If we measure the upper secondary terminal relative to the center tap, there is a 60 Hz sine wave with 14.1 V peak. On the other hand, the lower secondary terminal (relative to center tap) yields the same sine wave but is 180 degrees out of phase.

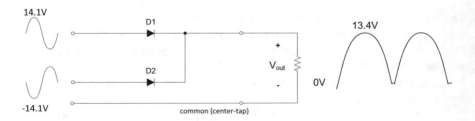

With the top terminal positive and the bottom terminal negative, D1 is on and D2 is off. Hence current flows from the input to the output with one diode drop, which gives an output peak 14.1 V − 0.7 V = 13.4 V. Similarly, if the top terminal is negative and the bottom terminal positive, D2 is on and D1 is off, and we get a rectified sine wave at the output.

Filtered Rectifiers

While the rectifier outputs have a non-zero DC average, the voltage varies significantly with time. To reduce this variation, we will use a capacitor to *filter* the rectifier output. To understand how a capacitor serves this purpose, let us look at some important facts about this passive device.

The voltage to current relationship for a capacitor is given by the equation below (Fig. 18):

$$i = C\frac{dv}{dt} = C\frac{\Delta v}{\Delta t}$$

The capacitor current depends on how fast the voltage is changing across the device. From this equation, we can draw a number of important conclusions:

- If the voltage across a capacitor does not change with time, then $i = 0$. In other words, a capacitor behaves like an open circuit under DC steady-state conditions.
- If the capacitor voltage changes with time, then its current is non-zero. Hence, the capacitor is no longer an open circuit but represents some "resistance" or *impedance* to the AC signal. For a sinusoidal voltage, the impedance of a capacitor is given by $1/(j2\pi f{\cdot}C)$, where f is the frequency of the sine wave.
- If a constant current flows into the capacitor, how does the capacitor voltage vary with time? Based on the equation, if $i = C{\cdot}dv/dt$ = constant, then dv/dt is also a constant. Since dv/dt is the slope of the capacitor voltage vs time graph, it means that the capacitor voltage increases linearly with time as seen below (Fig. 19).
- If the direction of the current source above is reversed, then the capacitor voltage will *decrease* linearly with time.
- The voltage across a capacitor cannot change instantaneously. To do so would imply that $dv/dt \to \infty$, and since $i = C{\cdot}dv/dt$, an infinite current would be required. The voltage across a capacitor may change quickly if a sufficiently large transient current is supplied.

Fig. 18 Capacitor voltage to current relationship $$i = C\frac{dv}{dt} = C\frac{\Delta v}{\Delta t}$$

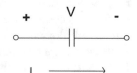

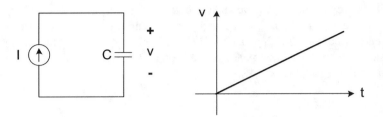

Fig. 19 Capacitor voltage vs time for a constant current

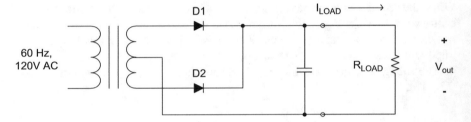

Fig. 20 Filtered rectifier circuit

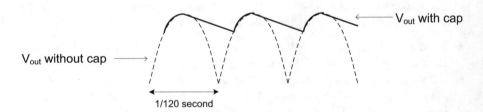

Fig. 21 Full wave rectifier output, with and without capacitor

With the capacitor factoids established, we can now look at how it is used in a filtered rectifier. Shown below is a full wave center-tapped rectifier with a capacitor connected to its output (Fig. 20):

For comparison, the rectifier outputs with and without a capacitor are shown below. Visually, we see that the "valleys" are being filled in, resulting in a smoother output voltage (Fig. 21).

A brief explanation of the circuit operation is provided in the diagram below. During every half cycle of the input (1/120 seconds in North America), current is supplied via the transformer and diodes to charge the capacitor; the capacitor voltage increases to V_{PK} as the secondary voltage reaches its peak. Typically, the duration for this charging takes only a small fraction of the 1/120 s period (Fig. 22).

As the secondary voltage passes the positive peak and begins to drop, the capacitor voltage does not change instantaneously and *holds* the output at this voltage. Both diodes are off, and the capacitor provides current to the load and discharges. In the next half cycle, the capacitor is replenished and charged up to V_{PK} again, and the process repeats.

Fig. 22 Filtered rectifier operation

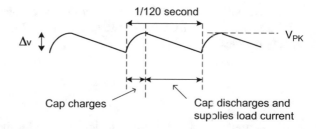

Fig. 23 Filtered rectifier load current approximation

$$i = C \frac{\Delta v}{\Delta t} = I_{LOAD}$$

$$\Delta v = \frac{I_{LOAD}}{C} \cdot \Delta t$$

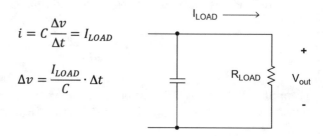

Ripple Voltage Estimate

From the description so far, the filtered rectifier output is not constant over time, but still has some residual ripple Δv. In this section, we will examine how the capacitor value affects Δv. We will make a series of approximations to derive an equation that estimates Δv for a given capacitance C.

Approximation #1 The load current I_{LOAD} is constant versus time: while obviously not true since the load voltage is equal to the capacitor voltage, if the ripple is small, this is not a bad approximation (Fig. 23).

$$i = C \frac{\Delta v}{\Delta t} = I_{LOAD.}$$

$$\Delta v = \frac{I_{LOAD}}{C} \Delta t$$

With this assumption, we can use the capacitor equation and set $i = I_{LOAD}$ and then rearrange the equation to solve for the ripple Δv.

Approximation #2 The capacitor charging time is much smaller than the discharging time. In other words, the discharge time Δt is close to 1/120 seconds, and by approximating $\Delta t = 1/120$:

$$\Delta v = \frac{I_{LOAD}}{C} \frac{1}{120} = \frac{I_{LOAD}}{Cf}$$

where $f = 120$ Hz for a full wave rectifier in North America

If the ripple is small, then the rectifier output voltage is approximately constant with a value of V_{PK}. By Ohm's law, $I_{LOAD} = V_{PK} / I_{LOAD}$. Substituting this into the equation yields

$$\Delta v = \frac{V_{PK}}{CfR_{LOAD}}$$

If the ripple is not small, then a better estimate for the rectifier output average is $V_{PK} - \Delta v/2$. Taking this into account, a modified equation for the ripple is given by

$$\Delta v = \frac{V_{PK}}{CfR_{LOAD} + 0.5}$$

Example Given a 20VCT transformer and the rectifier circuit below, estimate the output DC value and the ripple for $C = 470\ \mu F$ and $R_{LOAD} = 1\ k\Omega$

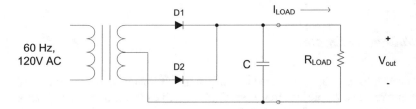

Answer Using results from previous examples, the half secondary voltage is $10V_{RMS}$, and the unfiltered rectifier output is $13.4\ V_{PK}$.

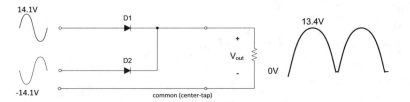

Assuming that the ripple is small, the output DC voltage is 13.4 V, resulting in a load current $I_{LOAD} = 13.4\ V/1\ k\Omega = 13.4\ mA$. Substituting the known values into the equation, we get $\Delta v = 13.4\ V/(470\ \mu F \cdot 120\ Hz \cdot 1\ k\Omega) = 0.24\ V$. The output DC voltage is approximately 13.4 V.

Example For the previous circuit with a 20VCT transformer, estimate the output DC value and the ripple for $C = 220\ \mu F$ and $R_{LOAD} = 1\ k\Omega$.

Answer Assuming that the ripple is small, the output DC voltage is still 13.4 V and the ripple has increased to 0.51 V, due to the lower capacitance value.

Is 0.51 V a small ripple? We can use the modified equation, which yields $\Delta v = 13.4 \text{ V}/(470 \text{ μF} \cdot 120 \text{ Hz} \cdot 1 \text{ kΩ} + 0.5) = 0.50 \text{ V}$. Since this result is not much different than before, we can say that the ripple is small

Example For the previous circuit with a 20VCT transformer, estimate the output DC value and the ripple for $C = 220 \text{ μF}$ and $R_{LOAD} = 100 \text{ Ω}$.

Answer Assuming that the ripple is small, the output DC voltage is still 13.4 V and the ripple has increased to 5.1 V, due to the higher load current discharging the capacitor faster. This ripple does not seem small, and the modified equation gives a ripple of 4.27 V. Hence, there is a discrepancy, and our assumption of small ripple is not a good one.

To that end, we need to also adjust the calculation of the output DC voltage. Referring to the diagram below, when the ripple is large, the average is given more accurately by $V_{PK} - \Delta v/2$.

In summary, V_{OUT} ripple $\Delta v = 4.27$ V, and the DC voltage is $13.4 \text{ V} - 4.27 \text{ V}/2 = 11.3 \text{ V}$.

5 Problems

1. Find the voltages and currents (V_D, V_R, I_D) in the circuit below:

   (a) $V_S = +10$ V
   (b) $V_S = -10$ V
   (c) $V_S = +0.2$ V

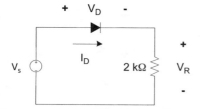

2. Find the voltages and currents (V_D, V_R, I_D) in the circuit below:

   (a) $V_S = +20$ V
   (b) $V_S = -20$ V
   (c) $V_S = +1$ V

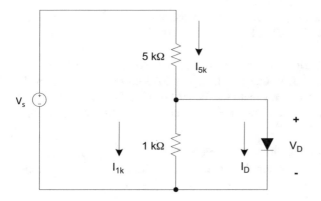

3. Find all mesh currents as indicated below:

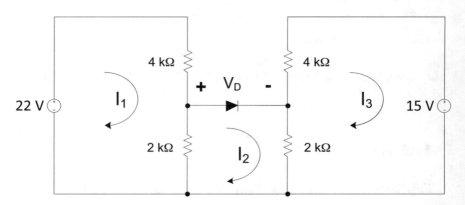

4. Given a 12 V transformer, sketch the rectifier output. Label both the voltage and time scale in your sketch.

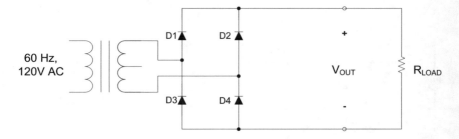

5. Given a 12VCT transformer, sketch the rectifier output. Label both the voltage and time scale in your sketch.

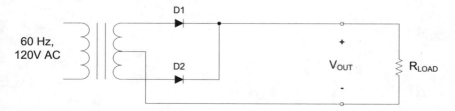

6. A 20VCT transformer is used below. Find the DC average and peak-to-peak ripple on V_{OUT}.

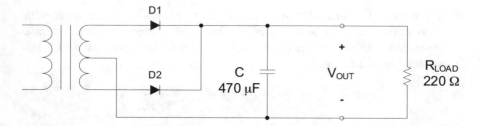

7. In the previous question, suppose that you need to reduce the ripple to 0.1 V_{pp}. What capacitor value is required?

8. Offline rectifiers connect directly to the line voltage to generate a rectified output. Assuming in North America that a 120 V AC is applied to input of a bridge rectifier, what is the maximum reverse bias voltage across the diode(s)? When choosing a diode, what reverse breakdown specification should be used?

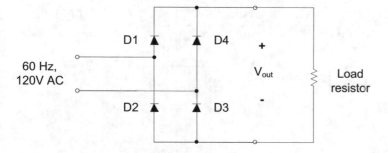

9. Find the voltage across D2 and the current through D3.

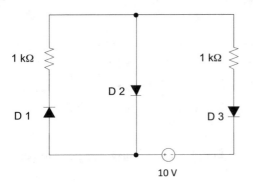

10. Tinkercad®: Create a half wave rectifier with a 1 kΩ load. A function generator with a 60 Hz, 5 V amplitude, 0 offset sine wave is connected to the rectifier input. Measure the rectifier output with a scope. The components used in this exercise are shown below.

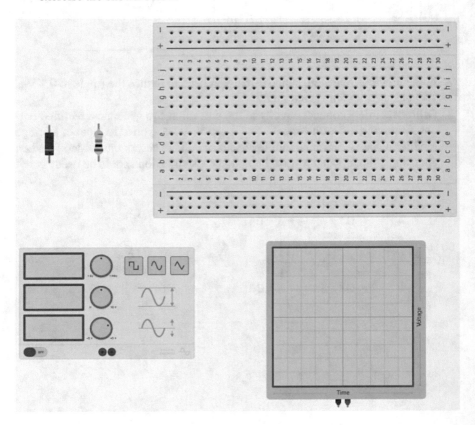

11. Tinkercad®: Using two function generators, connect them together to form a differential output. A differential output has a positive output and negative output, where the negative output is 180° out of phase relative to the positive output. Set both function generators to output a 60 Hz, 5 V amplitude, 0 offset sine wave.

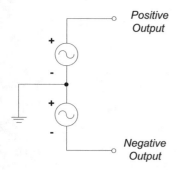

12. Tinkercad®: The differential output from the previous question is used to emulate a transformer secondary. Build a full wave center-tapped rectifier with a 1 kΩ load. The differential signals are connected to the rectifier input. Measure the rectifier output with a scope.

Semiconductor Physics

Before we move onto the study of transistors, we need to introduce some terminologies from semiconductor physics. While most of us have an intuitive idea about what conductors and insulators do, semiconductors appear to be more exotic. Do they "sort of" conduct? Pure elemental semiconductor is not very conductive, but we can alter its electrical properties in a controlled manner, and this is what makes semiconductors useful for building electronic devices.

Looking at the periodic table, you will recall that elements are grouped by the number of valence electrons within an atom. As shown below, Group III elements (e.g., boron) have three valence electrons, Group IV elements (e.g., carbon) have four valence electrons, and so forth. The semiconductors silicon and germanium fall under Group IV (Fig. 1).

The most common semiconductor material, silicon, has four valence electrons. In the diagram below, the left picture shows a single silicon atom, and each line is used to represent a valence electron. As we bring other silicon atoms into close proximity with each other, they can share valence electrons to form a *covalent* bond. With four valence electrons, each silicon atom forms bonds with four neighboring atoms, arranging themselves into a crystalline structure with 5×10^{22} atoms/cm^3 (Fig. 2).

At a temperature of absolute zero, all of the electrons would stay in place within the covalent bonds, and no charge would be free to wander around the crystal. As the temperature is raised, some electrons would gain enough energy to break free, allowing them to be mobile charge carriers. In semiconductor physics, this is described with an energy band diagram, which is shown below to the right. E_v and E_c depict the valence band energy and conduction band energy, respectively. We will not describe what these are right now, but the key idea is that for an electron to break free from a covalent bond, it must gain enough energy to move up to the conduction band E_c (Fig. 3).

In pure silicon at room temperature, there are approximately 1.5×10^{10} free electrons per cm^3. These free electrons, or carriers, are what makes the conduction of

© Springer Nature Switzerland AG 2022
C. Siu, *Electronic Devices, Circuits, and Applications*,
https://doi.org/10.1007/978-3-030-80538-8_3

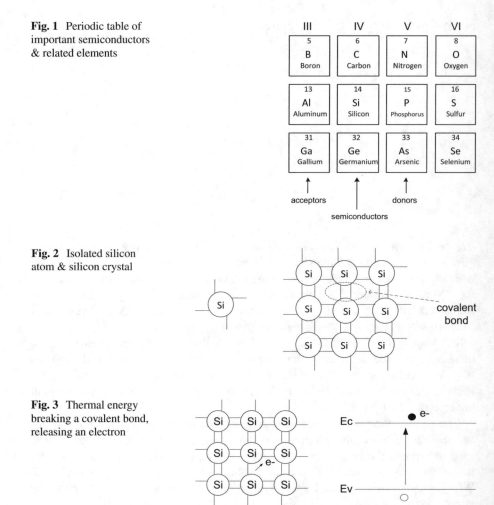

Fig. 1 Periodic table of important semiconductors & related elements

Fig. 2 Isolated silicon atom & silicon crystal

Fig. 3 Thermal energy breaking a covalent bond, releasing an electron

electrical current possible. Although that sounds like a lot of carriers, it is really a miniscule number compared to the atoms per cm^3; pure silicon is hence not very conductive. What makes silicon useful, aside from being sand on the beach, is that we can introduce certain impurities to change its conductivity.

1   N-Type Silicon

Phosphorus is a Group V element with five valence electrons. Recall that in pure silicon, the silicon atoms arrange themselves into a crystalline structure with four neighboring atoms. If we replace one of the silicon atoms with a phosphorus atom, it would form four covalent bonds with the neighbors, but the fifth valence electron

Fig. 4 Adding donor atoms to create N-type silicon

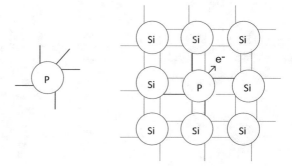

Fig. 5 Ionized donor atom in N-type silicon

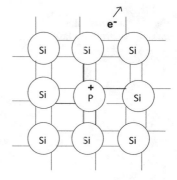

is not bonded and is free to move about. Hence, Group V elements act as donors of electrons to create *n-type* silicon, and by adjusting the amount of phosphorus, we can control the number of free electrons and the silicon conductivity. In semiconductor speak, we are *doping* the silicon with impurities (phosphorus) (Fig. 4).

Note also that since a valence electron has left the phosphorus atom, the phosphorus atom has a net positive charge that cannot move since the atom is fixed in place within the crystal (Fig. 5).

2 P-Type Silicon

In a similar manner, by introducing Group III atoms such as boron into pure silicon, the three valence electrons are not sufficient to complete the covalent bonds with the four neighboring silicon atoms. Hence, the boron atom will steal an electron from some other bond to complete all four neighboring covalent bonds (Fig. 6).

The bond that was robbed of an electron now has a net positive charge at that site. As we will see shortly, this positive charge can move around, and hence we can visualize it as a free carrier as shown below. In semiconductor physics, a positive charge carrier is known as a hole. Note that the boron atom, in accepting an electron,

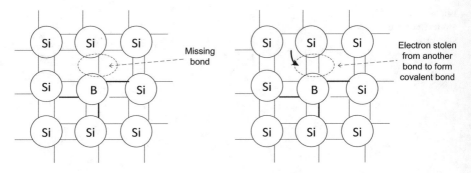

Fig. 6 Adding acceptor atoms to create P-type silicon

Fig. 7 Visualizing the
missing electron as a hole

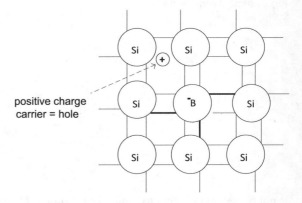

now has a negative charge associated with it. This negative charge cannot move
since the boron atom is fixed in place within the silicon crystal (Fig. 7).

The movement of the *hole* is shown in the figure below, as the missing bond
(hole) can steal an electron from another bond. A different bond at a different location is now missing an electron; hence it now has a net positive charge. If we focus
on the location of the hole, it seems as if the positive charge has moved to the right.
Hence, this is the reason why Group III elements act as acceptors of electrons to
create *p-type* silicon, in which the free carriers are positive charges or *holes* (Fig. 8).

In summary, some of the key facts regarding n-type and p-type silicon are summarized below (Table 1).

Semiconductor devices are constructed by combining n-type and p-type silicon
in a prescribed manner. The doping level, or amount of impurities, can also be
changed to adjust the device properties. As an example, a diode is made by joining
p-type silicon with n-type silicon, creating what is referred to as a P-N junction
(Fig. 9).

We will study the physics of a P-N junction later on, but based on earlier study
with the simple model, we know that a P-N junction only conducts current in one
direction.

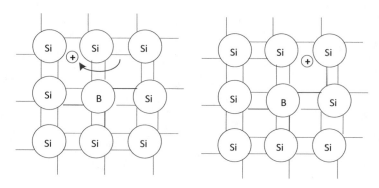

Fig. 8 Hole movement in P-type silicon

Table 1 N-type vs P-type silicon comparison

	Impurities added	Charge carriers
N-type silicon	Group V element = donor	Negative = electron
P-type silicon	Group III element = acceptor	Positive = hole

Fig. 9 The construction of a silicon diode

3 Problems

1. How many valence electrons are associated with one atom of a semiconductor such as silicon?
2. Is pure silicon at room temperature a good conductor? Why or why not?
3. What is a donor atom? How is it used in the manufacturing of semiconductor devices?
4. In semiconductor physics, what is a hole?
5. What is an acceptor atom? How is it used in the manufacturing of semiconductor devices?
6. How is a silicon diode constructed?

Introduction to the BJT

The ability to amplify electrical signals is a relatively recent discovery, starting with the invention of the vacuum tube in the early twentieth century. The vacuum tube, however, is large and power hungry; it works by heating a filament to high temperatures, similar to incandescent light bulbs. The search was on for a device that can perform electronic amplification more efficiently, and this culminated with the successful demonstration of the first transistor at Bell Labs in New Jersey, USA, in 1947. John Bardeen, Walter Brattain, and William Shockley won the 1956 Nobel Prize in physics for their invention.

The bipolar junction transistor (BJT) was the workhorse of the electronics industry until the 1990s, and even today, this transistor is used due to engineers' familiarity with it. The body of work associated with the BJT is so large that it warrants study by all circuit designers.

1 BJT Symbols and Terminologies

A BJT is created by sandwiching p-type and n-type silicon in a prescribed manner. We will not study the device physics of a BJT in this introduction, but instead focus on the terminal characteristics of a BJT and how to use them.

There are two flavors of BJT that we will contend with: the NPN and the PNP transistor. To begin, we will focus only on the NPN transistor, starting with the symbol and terminal names for this device (Fig. 1).

The construction of the NPN transistor follows its name: it starts with n-type silicon followed by p-type silicon, with a third layer made of n-type silicon again. There is a connection made to each of the three silicon pieces, giving the three terminals named *collector*, *base*, and *emitter*.

Given that a diode is made up of a P-N junction, and there are two P-N junctions in diagram (b), one may draw an equivalent circuit for the NPN as two diodes as in diagram (c). However in diagram (c), it would not be possible to conduct current

© Springer Nature Switzerland AG 2022
C. Siu, *Electronic Devices, Circuits, and Applications*,
https://doi.org/10.1007/978-3-030-80538-8_4

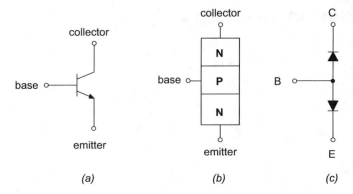

Fig. 1 NPN transistor (**a**) symbol (**b**) construction (**c**) conceptual diagram

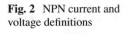

Fig. 2 NPN current and voltage definitions

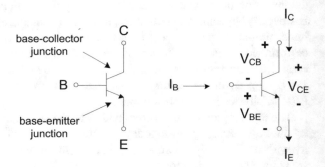

from collector to emitter; there are two back-to-back diodes in series from collector to emitter. Indeed, if you took two discrete diodes and connect them as shown above, it will not conduct current. What makes the operation of a NPN transistor possible is some special engineering when the device is manufactured (for those who wants to know, the P region in a NPN is very thin). This not only makes it possible for collector-to-emitter conduction, but we can control this current using the base terminal.

Before we learn how to use a BJT, there are some additional terminologies that need to be introduced (Fig. 2).

- As mentioned previously, the NPN is composed of two P-N junctions, which we will refer to as the base-collector (B-C) junction and the base-emitter (B-E) junction

- When placed into a circuit, the DC voltages of a BJT are defined using the nomenclature shown. For example, the voltage difference between the collector and emitter terminals is denoted as V_{CE}
- The conventional current flow for a NPN is shown as I_B, I_C, and I_E above; by Kirchhoff's current law (KCL), we can write the following:

$$I_E = I_C + I_B$$

2 NPN Transistor Operating Regions

To understand the operation and usage of a NPN, we can draw from our knowledge of diodes. Recall that using the simple model of 0.7 V, we can quickly determine whether the diode is on or off for different values of V_B (Fig. 3).

To determine whether the NPN is on, we need to check whether the B-E junction diode is on. As shown below, we can use the same diode techniques to make this analysis (Fig. 4).

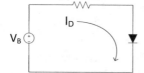

V_B (V)	Diode
0	Off
0.5	Off
1	On
2	On

Fig. 3 Silicon diode operation for various applied voltages

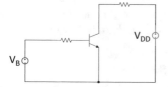

V_B(V)	B-E Diode	NPN
0	Off	Off
0.5	Off	Off
1	On	On
2	On	On

Fig. 4 NPN transistor operation for various applied voltages

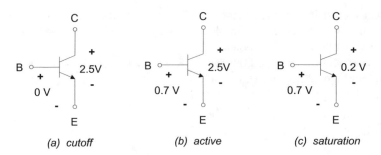

Fig. 5 NPN transistor DC operating regions

Key Point NPN is on if $V_{BE} = 0.7$ V

Exercise Calculate the DC base current I_B in the circuit below.

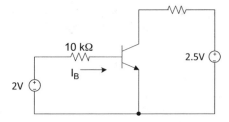

Answer Clearly, the 2 V connected to the 10 kΩ resistor is enough for $V_{BE} = 0.7$ V and turns on the NPN. The net voltage across the 10 kΩ resistor is 2 V – 0.7 V = 1.3 V, and $I_B = 1.3$ V/10 kΩ = 0.13 mA.

When the NPN is on, its behavior is also affected by the collector-emitter voltage V_{CE}. A transistor can be used as an electronic switch or an amplifier, and to that end, it must be biased in the correct DC operating region. The introduction of the BJT's operating regions is done using examples in the diagram below (Fig. 5).

- *Cutoff* – When $V_{BE} = 0$ V, B-E diode is off and the NPN is off. In general, V_{CE} can be any positive value when the NPN is off.
- *Active* – When $V_{BE} = 0.7$ V, the NPN is on and now we must examine V_{CE} to determine which region it is in. If $V_{CE} > 0.2$ V, the NPN transistor is in the active region. In diagram (b), $V_{CE} = 2.5$ V, and hence this NPN is in the active region and can be used as an amplifier. Note that this region is also known as *forward active* region or the *linear* region in literature.
- *Saturation* – If the NPN is on but V_{CE} has dropped to 0.2 V or less, we will make a simplification and set $V_{CE} = 0.2$ V. The NPN is in the saturation region and can be used as a closed switch, which is shown in diagram (c).

Key Point We define the saturation voltage $V_{CE(SAT)} = 0.2$ V. If a NPN has $V_{CE} > V_{CE(SAT)}$, the transistor is in the active region, otherwise it is in the saturation region.

Fig. 6 Relationship
between NPN currents in
the active region

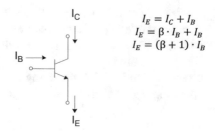

$$I_E = I_C + I_B$$
$$I_E = \beta \cdot I_B + I_B$$
$$I_E = (\beta + 1) \cdot I_B$$

3   DC Current Gain

We had mentioned that in the active region, the NPN can be used as an amplifier.
Intuitively, the transistor must be providing some gain to the electrical signal, and
for a BJT, one of the ways this is quantified is with the DC current gain parameter.
In the *active* region, there is a relationship between the collector current and base
current, given by

$$I_C = \beta \cdot I_B$$

where β = DC current gain = I_C / I_B. If we view the base terminal as the input and
collector as the output, then a small change in the input current will cause a large
change in the output. In datasheets, you will also see β represented as the parameter
h_{FE}; they both refer to the same thing.

Previously, we used KCL to deduce that $I_E = I_C + I_B$. In the active region, we can
use the β definition to see the relationship between I_E and I_B (Fig. 6).

To summarize, here are the equations for the BJT so far:

$I_E = I_C + I_B$	valid for all operating regions
$I_C = \beta \cdot I_B$	active region only
$I_E = (\beta + 1) \cdot I_B$	active region only

One of the important tasks in BJT circuit analysis is to estimate the DC operating
point (DCOP) of the transistor, which typically consists of the DC collector current
and collector-emitter voltage (I_C, V_{CE}). From this, we can establish which operating
region the transistor is in.

Exercise Find the DC operating point of the NPN below. Identify the transistor's
operating region. Assume that $\beta = 100$.

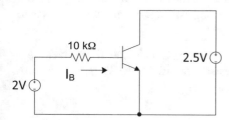

Answer In a previous example, we found that the NPN is on with $I_B = 0.13$ mA. V_{CE} is at 2.5 V due to the voltage source, and hence the transistor is in the active region with $I_C = 100 \cdot I_B = 13$ mA. The DCOP is (13 mA, 2.5 V).

Exercise Find the DC operating point of the NPN below. Identify the transistor's operating region. Assume that $\beta = 100$.

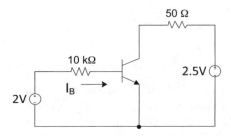

Answer Similar to the previous example, $I_B = 0.13$ mA, but V_{CE} is not 2.5 V due to the voltage drop across the 50 Ω resistor. The approach we will take is to *assume* that the NPN is in active region; then *check* if this assumption is correct or not.

With this assumption, $I_C = 100 \cdot I_B = 13$ mA, which gives $V_{CE} = 2.5$ V $-$ (50 Ω) (13 mA) $= 1.85$ V. Since $V_{CE} > 0.2$ V, our assumption is correct, and the DCOP $= (13$ mA, 1.85 V).

Exercise Find the DC operating point of the NPN below. Identify the transistor's operating region. Assume that $\beta = 100$.

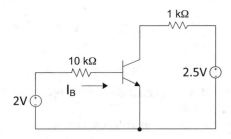

Answer $I_B = 0.13$ mA, and with the active region assumption, $I_C = 13$ mA. $V_{CE} = 2.5$ V $-$ (1 kΩ)(13 mA) $= -10.5$ V. Generally with only positive voltage sources in the circuit, no circuit node can have a negative voltage. Hence, V_{CE} being negative means that we made the wrong assumption, and the NPN is actually in saturation.

For a NPN in saturation, $V_{CE} = 0.2$ V, and now we can calculate the collector current by finding the current through the 1kΩ resistor. $I_C = (2.5$ V $- 0.2$ V$)/1$k$\Omega = 2.3$ mA, and the DCOP $= (2.3$ mA, 0.2V).

4   BJT I-V Curves

Just like with diodes, we can show the characteristics of a BJT using I-V curves. In particular, we can plot the collector current I_C vs the collector-emitter voltage V_{CE}. Recall that the slope of an I-V curve is related to the resistance of the device. What is the slope of the BJT I-V curve in the different operating regions?

For a BJT in the cutoff region, the transistor is off and conducts no current between collector to emitter. As such, the transistor looks like an open switch from collector to emitter, and the graph below shows zero collector current for various values of V_{CE} (Fig. 7).

For a BJT in the saturation region, the transistor is on and acts like a closed switch. In the model we have so far, we set $V_{CE} = 0.2$ V for saturation, but a more accurate model is to have a resistance R_{ON} between the collector and emitter. In an ideal closed switch, R_{ON} would be zero, but for a real device, there is a small finite resistance. Small resistances have steep slopes in I-V graphs, and hence we can represent the BJT between 0 and 0.2 V with the segment shown below (Fig. 8).

For a BJT in the active region, the transistor is on and can be used as an amplifier. At first it is not apparent how to represent an amplifier on an I-V graph, but let us start with the equation $I_C = \beta \cdot I_B$. The collector current is controlled by the base current, and hence the transistor is behaving as a current-controlled current source as depicted in Figure 9(a) below. If this dependent source is ideal, then the voltage V_{CE} does not affect the current, and it is represented as an ideal current source with a horizontal line on a I_C vs V_{CE} graph. Whereas the AC resistance of an ideal current source is infinite, in a real transistor the AC resistance is finite and given by r_o in

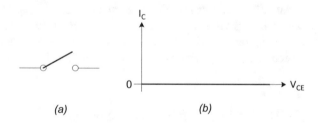

Fig. 7 NPN Cutoff Region, modeled as an Open Switch

(a)   (b)

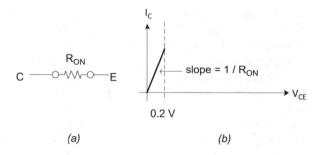

Fig. 8 NPN saturation region, modeled as a non-ideal closed switch

(a)   (b)

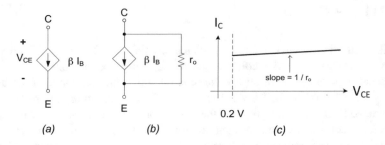

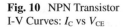

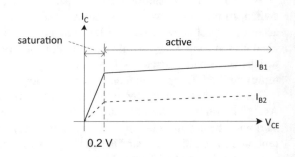

Fig. 9 NPN Active Region (**a**) Ideal Dependent Source (**b**) Improved Model (**c**) I-V Curve

Fig. 10 NPN Transistor
I-V Curves: I_C vs V_{CE}

diagram (b). Typically r_o is large, in the 10's to 100's of kΩ, and thus this is represented by a shallow slope in diagram (c) (Fig. 9).

Putting this all together, we have the I-V curve for a NPN transistor as shown below. The cutoff region is not shown since it is still a horizontal line at $I_C = 0$. A family of I-V curves is typically shown in literature, and two I-V curves for different base currents are shown where $I_{B1} > I_{B2}$. One way to think about this is that in the active region, $I_C = \beta \cdot I_B$, and so I_{B1} will produce a higher collector current than I_{B2}, which is precisely what is shown by the graph (Fig. 10).

5   Designing an Electronic Switch

The first application we will study is using a BJT as an electronic switch. The advantage of using a BJT vs a diode as a switch is that we now have a separate control terminal for opening and closing the switch.

Recall that to use the BJT as a switch, the transistor must be either in the *cutoff* or the *saturation* region. Our job in the design is to guarantee that the BJT stays in these regions for a given load specification. Consider a system with a 5 V power supply. When the control input is 0 V, the NPN is in cutoff and since there is no voltage drop across the load, V_{CE} is equal to the supply voltage of 5 V. When the control input is 5 V, however, we need to size the base resistor such that the NPN goes into saturation for the given load (Fig. 11).

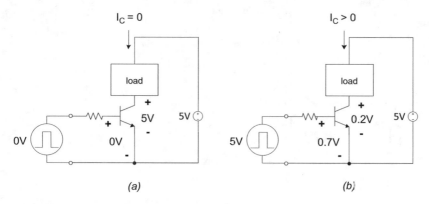

Fig. 11 NPN used as a Switch (a) switch off (b) switch on

Exercise The BJT below is being used as a switch. Estimate the load current if the switch is *on* for $V_{in} = 5$ V.

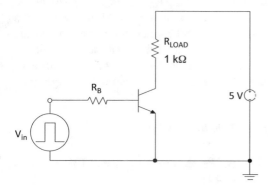

Answer The question states that the switch is on, implying that the NPN is in saturation. Since $V_{CE} = 0.2$ V, the current through the load is $(5\text{ V} - 0.2\text{ V})/1\text{ k}\Omega = 4.8$ mA, which is also equal to the collector current.

In the previous example, we need to choose R_B such that the NPN is guaranteed to be saturation when it is conducting the 4.8 mA current. The process of choosing R_B with some design margin or overdrive factor is illustrated by the next question.

Exercise A NPN has $\beta = h_{FE} = 50$. Design a switch for a 5 V system with an overdrive factor OD = 2.

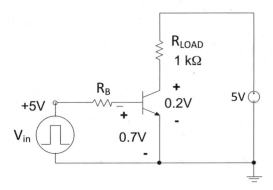

Answer The design process is broken down into several steps:

- *Step 1* – Use I_{LOAD} to find the base current $I_{B(MIN)}$. Since the NPN should be saturated, then $V_{CE} = 0.2$ V and $I_{LOAD} = I_C = 4.8$ mA. The minimum base current required to saturate the NPN, $I_{B(MIN)}$, is found by taking $I_{LOAD}/\beta = 4.8\,\text{mA}/50 = 96\,\mu\text{A}$.
- *Step 2* – Based on the overdrive factor (OD), determine the design target for the base current. The $I_{B(MIN)} = 96$ µA is the minimum base current needed to saturate the NPN, and we want some margin to ensure the NPN is saturated in spite of component tolerances, such as β variation from device to device. The overdrive factor is one way that margin is added by using a base current that is larger than $I_{B(MIN)}$.

$$I_B = \text{OD} \cdot I_{B(MIN)} = 2 \cdot 96\,\mu A = 192\,\mu A$$

- *Step 3* – Calculate the base resistor RB. We are targeting $I_B = 0.192$ mA, and hence $R_B = (5\text{ V} - 0.7\text{ V})/0.192\text{ mA} = 22.4$ kΩ.

DC Motor Driver

One of the uses for a BJT switch is DC motor control. Not only can we start and stop a DC motor with an electronic signal, but we can also control its speed.

A DC motor, as its name implies, operates using a DC power supply; it converts a DC voltage and current to mechanical energy. There are many resources on the Internet that explains how a DC motor works, so we will not duplicate it here. However, a few parts in the DC motor will be identified to bring context to the discussion that follows.

The traditional brushed DC motor consists of a stator and a rotor. A stator may consist of permanent magnets which do not move. The rotor, on the other hand, does rotate by passing current through the windings wrapped around this part. By changing the current direction at the right times, the motor can turn continuously (Fig. 12).

Fig. 12 Components of a
DC Motor

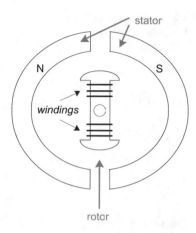

To implement an on/off control for the motor, we can use the NPN switch as shown below. Current must pass through the rotor windings to run the motor, and since the windings create an inductance, the NPN switch is effectively driving an inductive load. The symbol we will use to represent a DC motor is shown in figure (a), and the equivalent inductance of the motor is depicted in figure (b). Also, we will depict a DC power supply with the V_{DD} symbol shown in the schematics below (Fig. 13).

To understand the issues of driving an inductive load, let us review the behavior of an inductor. The voltage-to-current relationship of an inductor is shown below, and an important insight is that the current through an inductor cannot change instantaneously; to do so would require an infinite voltage across the inductor (Fig. 14).

When we turn the BJT on, current builds up in the DC motor, and energy is stored in the magnetic field of the rotor windings. However, the moment we shut the BJT off, this magnetic field will not disappear instantly, and the rotor inductance will attempt to keep the current going in the same direction. This is the physical reason why inductor current cannot change instantaneously (Fig. 15).

Figure 15 shows what happens at the moment the BJT is turned off. In (a) when the transistor is on, there is a current I_M through the motor that generates mechanical motion. At the moment that the input transitions to 0 V and the NPN shuts off, the motor current does not go to 0 right away due to the inductance. Figure (b) shows the current I_M through the motor, even though the NPN is off. Where does this current go?

One way to view the inductor is that it has stored energy, but after we shut the transistor off, it will release that energy. The inductor will act like a generator for a brief time, and the motor voltage reverses polarity to V_{RM} as a result. The voltage on the NPN collector has now increased to $V_{DD} + V_{RM}$, which can be well above the supply voltage! This voltage spike may cause the NPN collector to breakdown, and so we must provide some protection to the transistor when the motor is turned off.

Fig. 13 NPN DC Motor
Driver (**a**) circuit (**b**)
equivalent load

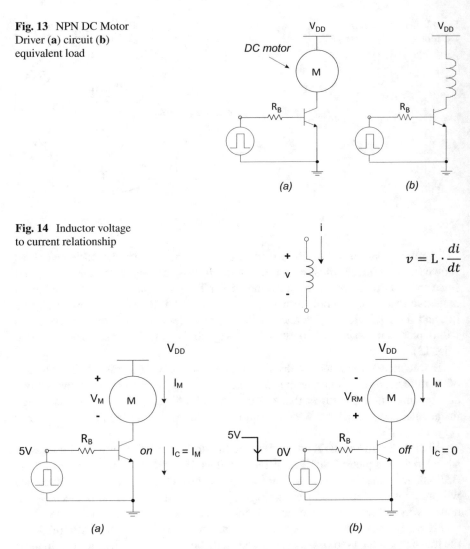

Fig. 14 Inductor voltage
to current relationship

Fig. 15 Motor driver (**a**) switch is on with motor running (**b**) the instant after switch turns off

Freewheeling Diode A diode can be added in parallel with the motor, such that when the motor is turned off, the current can be shunted through the diode, in effect clamping the collector voltage to a safe level; the collector voltage is $V_{DD} + 0.7$ V when the freewheeling diode is on (Fig. 16).

With a single transistor, we can implement motor on/off control, but the motor only runs in one direction. If we need to have the motor run in either direction, a different circuit topology is needed: the H-Bridge.

Fig. 16 DC motor driver
with freewheeling diode

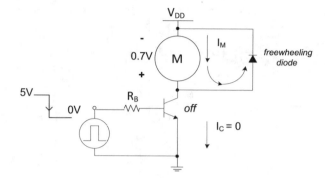

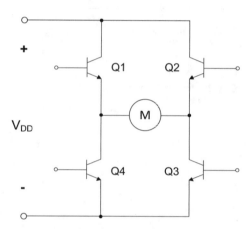

Fig. 17 H-Bridge circuit

The H-Bridge

The H-bridge consists of four transistors Q1 to Q4. Each transistor is used as a switch, being either on or off. Instead of assembling this using discrete transistors, there are integrated circuits available with the H-Bridge, such as the L293D that includes the freewheeling diodes on-chip (Fig. 17).

To understand how the H-Bridge works, we can refer to the figure below. If Q1 and Q3 are on but Q2 and Q4 are off, then current passes through the motor in one direction. If Q1 and Q3 are off but Q2 and Q4 are on, then current goes in the opposite direction. This way, we can control the motor to move forward or backward (Fig. 18).

6   Using the PNP Transistor

We now turn our attention to the application of the PNP transistor. The usage of the PNP is similar to the NPN, but some of the polarities need to be reversed. Let us begin with some definitions and terminologies for the PNP (Fig. 19).

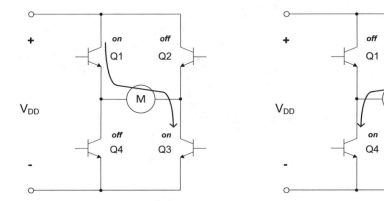

Fig. 18 H-Bridge operation

Fig. 19 PNP transistor (**a**) symbol (**b**) construction (**c**) conceptual diagram

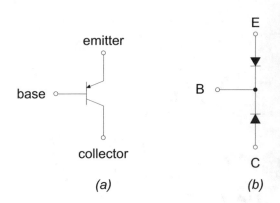

Note that the PNP symbol differs from the NPN symbol in one aspect: the arrow on the emitter terminal is reversed. However, the author has also chosen to draw the PNP symbol with the emitter terminal on top, and this is done for a reason that will be explained shortly.

To turn on the PNP, the emitter-base diode must be on, meaning that the emitter voltage is *higher* than the base voltage in a PNP. This is opposite to that of a NPN. In circuit design, it is common to find the NPN emitters tied to the lowest potential in a system (e.g., common), whereas the PNP emitters are tied to the highest potential (e.g., V_{DD}). Hence by drawing the PNP emitter at the top, the voltages go from highest to lowest as we scroll down on the schematic. This in turn makes the schematic more readable and easier to analyze (Fig. 20).

Using transistors as switches, we have a choice between a NPN or PNP: a switch connected to ground or to V_{DD}. These are called low-side switches and high-side switches, respectively (Fig. 21).

Low-side switches are more common than high-side switches, but there are reasons to use the latter type. For example, high-side switches are preferred for modern automotive uses. One of the reasons is that the body of a car is metal, and 95% of

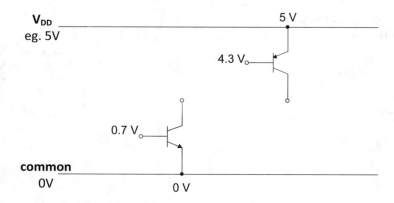

Fig. 20 Typical usage of NPN and PNP transistors

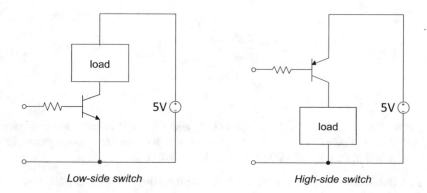

Fig. 21 Low side & high side switches

the total car is ground. Therefore, if the load is accidentally shorted to the car body and ground, an opened high-side switch will isolate the battery from the fault.

The equations for the PNP are the same as that for the NPN, with a couple of adjustments or notation changes. First, by KCL, the emitter current is still the sum of I_B and I_C. Note however that the base current flows out of the base terminal, which is the opposite of NPN (Fig. 22).

In the active region, $I_C = \beta \cdot I_B$ just as before. However, the emitter to collector voltage V_{EC} must be >0.2 V for the PNP to be in the active region; otherwise it is in saturation.

Example Find the DC operating point for the PNP transistor below. Assume that $V_{EB(ON)} = 0.7$ V and $V_{EC(SAT)} = 0.2$ V.

Fig. 22 Current flow in a
PNP transistor

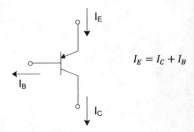

$$I_E = I_C + I_B$$

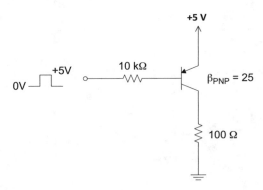

Answer When V_{in} = +5 V, then V_{EB} = 0 V and the PNP is off. However when V_{in} = 0 V, then V_{EB} = 0.7 V and the PNP is on. The base current flowing out of the base terminal is then I_B = (5 V – 0.7 V)/10 kΩ = 0.43 mA.

Just like in the NPN exercises, we now calculate the collector current assuming that the PNP is in the active region:

$$I_C = \beta \cdot I_B = 25 \cdot 0.43\,\text{mA} = 10.75\,\text{mA}$$

The resulting collector voltage is V_C = (10.75 mA)(100 Ω) = 1.075 V. The emitter-collector voltage is thus

$$V_{EC} = 5\,\text{V} - 1.075\,\text{V} = 3.925\,\text{V}$$

Since $V_{EC} > 0.2$ V, our assumption is correct and the PNP is in the active region. The DCOP is (10.8 mA, 3.93 V).

Example Find the DC operating point for Q1 and Q2.

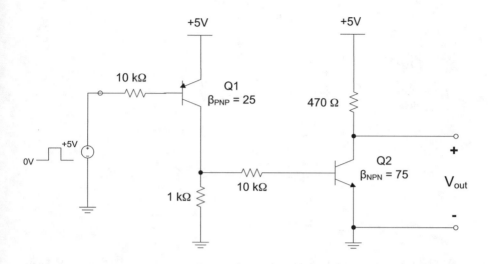

Answer We will find the DC operating point of Q1 first, followed by Q2. If $V_{in} = +5$ V, the Q1 PNP is off and the collector voltage $V_{C1} = 0$ V (the current through the 1 kΩ resistor is zero, hence the voltage drop is zero). As a result, the Q2 NPN is also off, and $V_{OUT} = +5$ V.

If $V_{in} = 0$ V, then the PNP Q1 is on, and its base current $I_{B1} = (5$ V $-$ 0.7 V)/10kΩ $= 0.43$ mA. Assuming that Q1 is in the active region, the collector current $I_{C1} = 25 \cdot 0.43$ mA $= 10.75$ mA, part of which goes into the 1kΩ resistor, and part of which goes to the base of Q2. To find the Q1 collector voltage V_{C1}, we can perform KCL on this node, and the reader is encouraged to do this as an exercise.

The simplified approach we will take is based on the observation that base currents are small compared to collector currents. Hence, if the Q2 base current I_{B2} is much smaller than I_{C1}, then most of I_{C1} will go through the 1 kΩ resistor. We can find the approximate value for V_{C1} by taking $I_{C1} \cdot 1$ kΩ, which yields 10.75 V. Note that the supply voltage is +5 V, and so it is impossible for an internal node in this circuit to be at 10.75 V. Our assumption that Q1 is in the active region is incorrect; it is actually in saturation. It is helpful to capture what we know so far and summarize it in the schematic below.

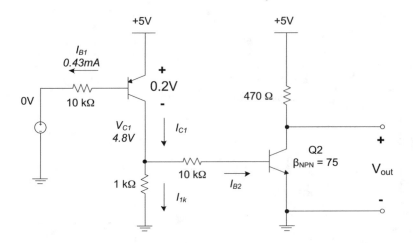

With Q1 in saturation, $V_{EC1} = 0.2$ V, and therefore $V_{C1} = 5$ V $- 0.2$ V $= 4.8$ V. From this, the 1 kΩ resistor current $I_{1k} = 4.8$ V/1 kΩ $= 4.8$ mA, and the Q2 base current $I_{B2} = (4.8$ V $- 0.7$ V$)/10$ kΩ $= 0.41$ mA. By KCL, $I_{C1} = I_{1k} + I_{B2} = 4.8$ mA $+ 0.41$ m A $= 5.21$ mA.

Now that we know what the Q2 base current is, we will assume that Q2 is in the active region and performs the usual procedure. $I_{C2} = (75)(0.41$ mA$) = 30.75$ mA. The Q2 collector voltage is $+5$ V $- (30.75$ mA$)(470$ Ω$) = -9.45$ V, and since this is not possible, Q2 is actually in saturation. $V_{out} = 0.2$ V, and $I_{C2} = (5$ V $- 0.2$ V$)/470$ Ω $= 10.2$ mA.

In summary, the DCOP for Q1 is (5.21 mA, 0.2 V), and the DCOP for Q2 is (10.2 mA, 0.2 V).

7   BJT Switch vs Ideal Switch

Having seen how a BJT can be used as a switch, how close is it to an ideal switch? Let us consider these two effects:

(a) In an ideal closed switch, the voltage drop is zero for any current through the switch. For a BJT in saturation, the voltage drop is 0.2 V. The current through the BJT multiplied by 0.2 V creates a power loss in the transistor (Fig. 23).
(b) To keep the BJT on, the control signal needs to continuously provide base current. In the diagram below, the control source dissipates 5 V·I_B to keep the BJT on (Fig. 24).

Both mechanisms cause power losses when using a BJT switch. If a large number of switches are needed, the power dissipation and heat can be significant. As we will see in the next chapter, there is another transistor that behaves closer to an ideal switch: the metal oxide semiconductor field effect transistor (MOSFET).

Fig. 23 NPN transistor in
saturation

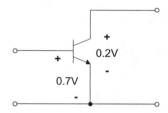

Fig. 24 Power supplied by
input source to keep NPN
transistor on

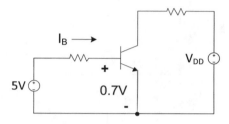

8 Problems

For all questions, assume that $V_{BE(ON)} = 0.7$ V, $V_{CE(SAT)} = 0.2$ V for the NPN transistor,
and $V_{EB(ON)} = 0.7$ V, $V_{EC(SAT)} = 0.2$ V for the PNP transistor.

1. An NPN transistor is in the common emitter configuration.

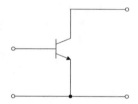

   Determine the region of operation (cutoff, active, saturation) for each of the
   following conditions:

	V_{BE} (V)	V_{CE} (V)	Operating region
(a)	0	5	
(b)	0.7	5	
(c)	0.5	5	
(d)	0.7	1	
(e)	0.7	0.2	

2. For each circuit below, determine the DC operating point (I_C, V_{CE}) and the
   region of operation. Assume that $\beta = 75$.

(a)

(a)

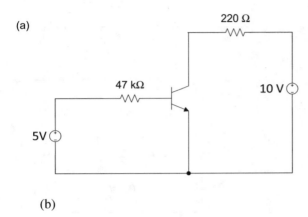

(b)

(b)

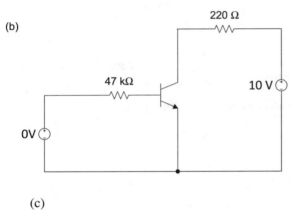

(c)

(c)

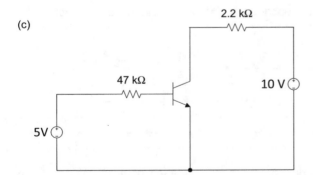

3. Design the BJT switch below by calculating a value for R_B using an overdrive factor of 1.5. Choose R_B out of 10% preferred values. Assume that $\beta = 10$.

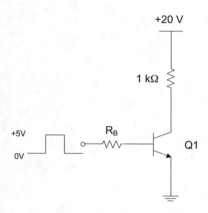

4. A high-side switch is required using a PNP transistor with $\beta = 30$. The maximum load current is 1 A while operating on a 12 V battery. The high-side switch is controlled by a 5 V logic signal. Design a switch that meets these requirements and sketch the schematic.

5. The circuit below is a classic topology used for biasing a BJT amplifier. Find the DC operating point for Q1.

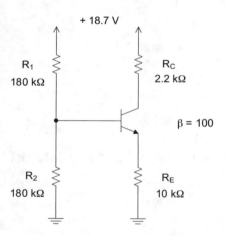

6. Calculate the DC operating points for Q1 and Q2, and find V_{OUT}.

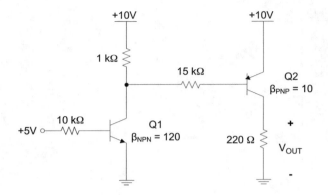

7. Calculate the DC operating points for Q1 and Q2.

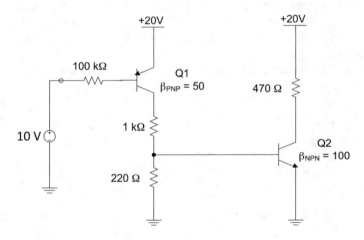

8. Design the switching circuit below with an overdrive factor of 2, by finding R1 and R2. Assume that $\beta = 200$.

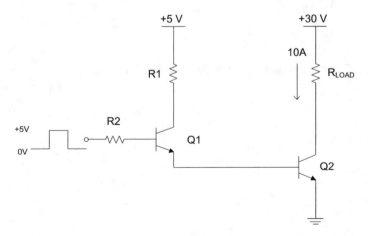

9. What is the purpose of a freewheeling diode?
10. The L293D is an integrated circuit with the following:

    - Two full H-bridges
    - Integrated freewheeling diodes
    - Interface circuitry for H-bridge control using logic signals

        Search for the L293D datasheet on the Internet and study it. Using the version in the 14-pin DIP package, sketch a schematic using the L293D such that:

    - The L293D is running on a +5 V power supply.
    - One of the H-bridges on the IC is used to drive a DC motor; the other H-bridge is not used.
    - One logic signal enables or disables the motor.
    - A second logic signal controls the direction of the motor when enabled.

11. Tinkercad®: Construct the following circuit on a virtual breadboard with a NPN transistor. After running the simulation, calculate beta from the ammeter readings.

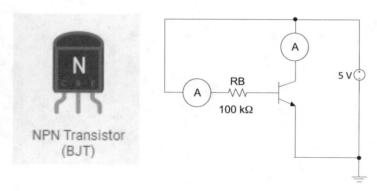

NPN Transistor
(BJT)

12. Tinkercad®: Using the NPN transistor, design a switch to support a 1A load on a +5 V supply. Use an overdrive factor of 1.5. Test your circuit by setting the input to 0 V and running the simulation. Then change the input to 5 V and run the simulation again.
13. Tinkercad®: Connect a 5 V power supply to a DC motor, and measure the current drawn by the motor.
14. Tinkercad®: Implement your L293D design from a previous question. Verify that you can enable and control the direction of a DC motor.

Introduction to MOSFETs

The MOSFET is the workhorse of the modern electronics industry. Since the BJT has been in use for longer, there exists a large catalog of circuits using the BJT. However, most integrated circuits designed and manufactured today use MOSFETs in complementary metal oxide semiconductor (CMOS) technology. There are several reasons behind the dominance of MOSFETs, one of which is its ability to approximate an ideal switch.

1 Characteristics of an Ideal Electronic Switch

An ideal electronic switch should exhibit the following behavior:

- When the switch is opened, there is an infinite resistance between its two terminals.
- When the switch is closed, there is zero resistance between its two terminals.
- The switch can be opened or closed using a voltage on a third terminal; no power is dissipated driving this third terminal (Fig. 1).

We will see that a MOSFET can satisfy these criteria reasonably well, or at least much better than a BJT can

2 MOSFET Symbols and Terminologies

Broadly speaking, there are two groups of MOSFETs: depletion mode and enhancement mode. Within each of these groups, there are PMOS and NMOS transistors. In this chapter, we will focus solely on enhancement-mode MOSFETs, which is the most common type found in circuit design.

© Springer Nature Switzerland AG 2022
C. Siu, *Electronic Devices, Circuits, and Applications*,
https://doi.org/10.1007/978-3-030-80538-8_5

Fig. 1 Ideal Electronic
Switch (**a**) opened (**b**)
closed

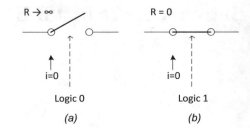

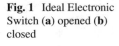

Fig. 2 NMOS symbol (**a**)
3-terminals (**b**) 4-terminals

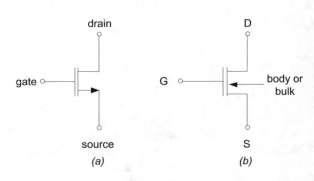

Shown below are schematic symbols for the NMOS transistor: there is a three-terminal symbol and a four-terminal symbol. Common to both symbols are the terminals gate, drain, and source. We will see that the gate is used to control the conductivity between drain and source, just like an electronic switch. The four-terminal symbol shows an additional, intrinsic connection called the *bulk* or *body*. In most discrete transistors, the *bulk* and *source* terminals are shorted together. In integrated circuit design, the bulk and source terminals may be connected separately, for reasons that we will skip over for now (Fig. 2).

The Threshold Voltage

Similar to the BJT, we can define the voltages across any two terminals of the NMOS as shown below (Fig. 3):

An important parameter of enhancement-mode MOSFETs is the *threshold voltage* V_t. By setting the NMOS's gate-to-source voltage (V_{GS}) above the threshold, we will turn the NMOS transistor on; otherwise it is off.

Key Point

If $V_{GS} > V_t$, then NMOS transistor is on.

Fig. 3 NMOS current and
voltage definitions

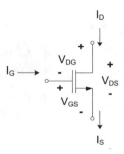

Exercise The NMOS below has a threshold voltage of 1.5 V. Determine whether
the transistor is on or off in each case.

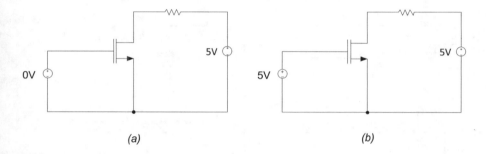

(a)   (b)

Answer (a) off, (b) on.

Construction of an NMOS Device

To understand some of the key characteristics and nomenclature behind the
MOSFET, it is helpful to see how a NMOS transistor is made. Transistors and integrated circuits are made on thin wafers of silicon, as shown in the Fig. 4. The largest
wafer being used today is 12″ in diameter, but less than 1 mm thick. A series of steps
are used to manufacture transistors in a silicon wafer.

- The entire silicon wafer is doped with acceptor atoms to create p-type silicon; an
  electrical connection can be made to this *p-type silicon substrate*, which results
  in the *body* or *bulk* terminal of a NMOS. See Fig. 5a.
- Donor atoms are implanted to create small n-type silicon "islands"; this is where
  the *source* and *drain* terminals of a NMOS are connected. See Fig. 5b.
- Silicon dioxide is grown on the surface of the silicon, in the areas between the
  source and drain "islands"; note that silicon dioxide is an insulator. See Fig. 5c.
- Finally, metal is deposited on top of the silicon dioxide, forming the *gate* terminal of a NMOS. See Fig. 5d.

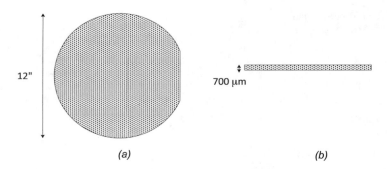

Fig. 4 Silicon wafer

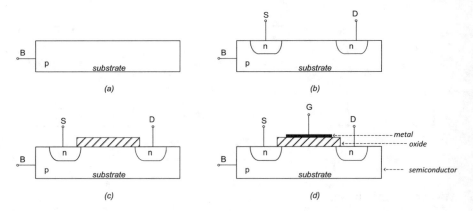

Fig. 5 Construction of a NMOS (**a**) body terminal connects to substrate (**b**) n-type islands connect to drain and source terminals (**c**) silicon dioxide is deposited (**d**) metal is put on top of oxide to form the gate terminal

3   NMOS Transistor Operating Regions

Recall that to turn an NMOS on or off, we must set the gate-to-source voltage VGS to either above or below the threshold voltage. If the NMOS is on, we need to make sure that it is in the correct operating region for our application; just like a BJT, the NMOS can be used as an amplifier or a switch. For this, we need to define an important quantity known as V_{DSAT}.

$$V_{DSAT} = V_{GS} - V_t$$

Key Points

- If $V_{DS} > V_{DSAT}$, the NMOS is used as an amplifier.
- If $V_{DS} < V_{DSAT}$, the NMOS is used as a closed switch.

Exercise   The NMOS below has a threshold voltage of 1.5 V. Determine whether it is biased for use as a switch or an amplifier.

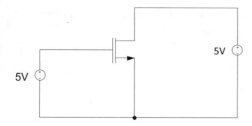

Answer $V_{GS} = 5$ V, the NMOS is on, and $V_{DSAT} = 5$ V $- 1.5$ V $= 3.5$ V. V_{DS} is also 5 V in this circuit, and since $V_{DS} > V_{DSAT}$, the NMOS is set up for amplifier uses.

Let us summarize what we have so far and also define the names of the MOSFET's DC operating regions: *cutoff*, *triode*, and *saturation* (Table 1).

While some of the names are the same as the BJTs, only the cutoff region has the same function. Unfortunately, saturation means different things for BJTs and MOSFETs, as we can see by comparison below (Table 2).

Using device physics, the equations of a MOSFET can be derived, but they are given here without all the leg work:

$$I_D = K\left(V_{GS} - V_t\right)^2 \qquad \text{NMOS in saturation}$$

$$I_D = 2K\left[\left(V_{GS} - V_t\right)V_{DS} - \frac{V_{DS}^2}{2}\right] \qquad \text{NMOS in triode}$$

where K and V_t are transistor parameters that can be derived from datasheets and will simply be given to us in this chapter. The following examples will show how these equations can be used in finding the DC operating point of a MOSFET.

Table 1 DC operating regions of a NMOS transistor

NMOS DC operating regions	V_{GS}	NMOS	V_{DS}
Cutoff	$< V_t$	Off	> 0
Triode	$> V_t$	On	$< V_{DSAT}$
Saturation	$> V_t$	On	$> V_{DSAT}$

Table 2 BJT vs MOSFET DC operating regions

BJT	MOSFET	Application
Cutoff	Cutoff	Open switch
Active	Saturation	Amplifier
Saturation	Triode	Closed switch

Example An NMOS transistor is biased in the saturation region. The transistor has the following parameters, biased at the following voltage(s):

- $K = 278$ mA/V^2.
- $V_{GS} = 4$ V.
- $V_t = 2$ V.

Find the drain current I_D.

Answer The example states that the NMOS is in saturation, so we can use the corresponding equation:

$$I_D = K\left(V_{GS} - V_t\right)^2 = \left(278\,\frac{\text{mA}}{\text{V}^2}\right)\left(4\,\text{V} - 2\,\text{V}\right)^2 = 1.11\,\text{A}.$$

Example Find the DC operating point of the NMOS in the circuit below, given the parameters

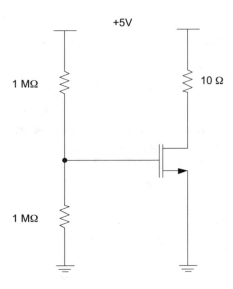

+5V

1 MΩ

10 Ω

1 MΩ

- $K = 0.2$ A/V^2.
- $V_t = 1.5$ V.

Answer An important thing to remember that the DC gate current is zero, so to calculate the gate voltage V_G, voltage division due to the 1 MΩ resistors gives 2.5 V.

With the source terminal connected to common, V_{GS} is also 2.5 V. Just like the approach we took to solve BJT DC operating points, we assume that the NMOS is in the saturation region (amplifier), perform the calculations, and then check to see if the assumption is valid.

$$I_D = K(V_{GS} - V_t)^2 = \left(0.2\frac{A}{V^2}\right)(2.5\,V - 1.5\,V)^2 = 0.2\,A$$

This in turn causes a voltage drop on the 10 Ω resistor connected to the drain. The drain-source voltage is

$$V_{DS} = 5\,V - (0.2\,A)(10\,\Omega) = 3V$$

We need to compare this against V_{DSAT}, which is

$$V_{DSAT} = 2.5\,V - 1.5\,V = 1\,V$$

Since $V_{DS} > V_{DSAT}$, the NMOS is in saturation and our assumption is correct. The DCOP of the NMOS is $(I_D, V_{DS}) = (0.2A, 3\,V)$.

Example Find the DC operating point of the NMOS in the circuit below, given the parameters:

- $K = 0.2\ \text{A/V}^2$.
- $V_t = 1.5$ V.

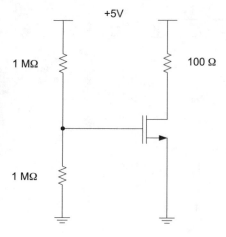

Answer The circuit is identical to that of the previous example, except that the drain resistor is now 100 Ω. Assuming that the NMOS is in saturation and using the previous results, $V_{GS} = 2.5$ V, $I_D = 0.2$ A, and $V_{DSAT} = 1$ V. The drain-source voltage V_{DS} is now

$$V_{DS} = 5\,V - (0.2\,A)(100\,\Omega) = -15\,V$$

Since it is impossible to have negative node voltages in a circuit running off a single positive power supply, we made the wrong assumption: the NMOS is actually in the *triode* region. To find the DC operating point, we must use the other MOSFET device equation, and we will demonstrate two methods of using this equation: the approximate method and the exact method.

Approximate Method in the Triode Region

Starting with the triode equation, we can make a simplification if V_{DS} is small. If $V_{DS} \ll 1$, then V_{DS}^2 is even smaller, and we can drop the squared term from the equation

$$I_D = 2K\left[\left(V_{GS} - V_t\right) \cdot V_{DS} - \frac{V_{DS}^2}{2}\right] \cong 2K\left[\left(V_{GS} - V_t\right) \cdot V_{DS}\right]$$

Taking this equation and forming the ratio V_{DS}/I_D, we can get the equivalent resistance seen between the drain and source

$$R_{DS(ON)} = \frac{V_{DS}}{I_D} = \frac{1}{2K\left(V_{GS} - V_t\right)}$$

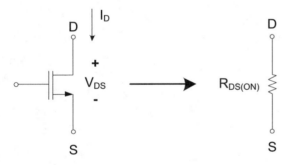

A MOSFET in triode acts like a resistance between the drain and source. Since we use the MOSFETs as switches in this operating region, we want $R_{DS(ON)}$ to be zero ideally. Real devices will not have zero resistances, but the $R_{DS(ON)}$ equation provides some guidance on how to lower this resistance. One way is to increase the gate-source voltage V_{GS}, while another way is to increase the mysterious transistor parameter K. Transistor manufacturers can make tradeoffs and increase K, with modern power MOSFETs having $R_{DS(ON)}$ of only a few milli-Ohms.

For this example, the NMOS on-resistance is

$$R_{DS(ON)} = \frac{V_{DS}}{I_D} = \frac{1}{2(0.2\,A/V^2)(2.5\,V - 1.5\,V)} = 2.5\,\Omega$$

If we replace the NMOS with this equivalent resistance between drain and source, we will get the following circuit:

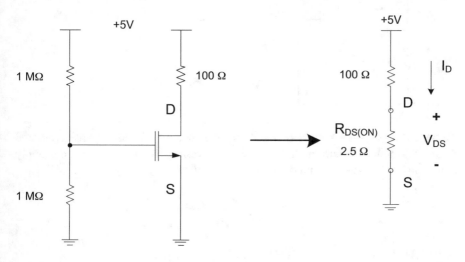

Applying Ohm's law and voltage division to the equivalent circuit, we can obtain $I_D = 48.8$ mA and $V_{DS} = 0.122$ V.

This idea of viewing the MOSFET in triode as a resistor is an important one. Note that we had assumed that V_{DS} is small in this approximation, and so is $V_{DS} = 0.122$ V small enough? In the second part of the solution, we will solve V_{DS} exactly and compare.

Exact Method in the Triode Region

In this method, we will retain the V_{DS}^2 term in the MOSFET triode equation

$$I_D = 2K\left[(V_{GS} - V_t)\cdot V_{DS} - \frac{V_{DS}^2}{2} \right]$$

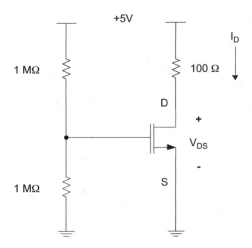

We have one equation with two unknowns in I_D and V_{DS}; K and V_t are given, and $V_{GS} = 2.5$ V. Since the current through the 100 Ω resistor is equal to I_D, we can apply Ohm's law to this resistor and obtain a second equation

$$I_D = \frac{5\,V - V_{DS}}{100\,\Omega}$$

Equating the expressions above, we get

$$\frac{5\,V - V_{DS}}{100\,\Omega} = 2K\left[(V_{GS} - V_t)\cdot V_{DS} - \frac{V_{DS}^2}{2}\right]$$

Substituting the known quantities and simplifying, we get a quadratic equation in V_{DS},

$$\frac{5\,V - V_{DS}}{100\,\Omega} = 2\left(0.2\frac{A}{V^2}\right)\left[(2.5\,V - 1.5\,V)\cdot V_{DS} - \frac{V_{DS}^2}{2}\right]$$

$$20\cdot V_{DS}^2 - 41\cdot V_{DS} + 5 = 0$$

The two solutions to the quadratic are $V_{DS} = 1.92$ V and 0.130 V. To determine which solution(s) is valid, we need to remember that NMOS is in triode with $V_{DSAT} = 2.5$ V $- 1.5$ V $= 1$ V. The first solution to V_{DS} would put the NMOS in saturation, which does not match our earlier findings. The second solution of $V_{DS} = 0.130$ V corresponds to the NMOS in triode, and so this is the valid solution.

Lastly to find I_D, we can substitute V_{DS} back into one of the equations and get $I_D = (5$ V $- 0.130$ V$)/100$ $\Omega = 48.7$ mA. The NMOS DCOP is thus 48.7 mA, 0.130 V. We see that the approximate solution is very close to the exact solution.

4   MOSFET I-V Curves

For a MOSFET in the cutoff region, the transistor is off and conducts no current between drain to source. It thus looks like an open switch, and the drain current is zero for various values of V_{DS} (Fig. 6).

For a MOSFET in triode, the transistor is on and behaves like a closed switch. However, the switch is not ideal, and there is a resistance R_{ON} between the drain and source. In last section's examples, we created a formula to calculate R_{ON}; for a given $V_{GS1} > V_t$, $R_{DS(ON)}$ is estimated by

$$R_{DS(ON)} = \frac{V_{DS}}{I_D} = \frac{1}{2K\left(V_{GS1} - V_t\right)}$$

We will assume that the I-V curve is linear over the triode region until $V_{DS} = V_{DSAT}$, when the transistor exits triode and enters saturation (Fig. 7).

For a MOSFET in the saturation region, the transistor is on and can be used as an amplifier. Recall the NMOS saturation equation:

$$I_D = K\left(V_{GS} - V_t\right)^2$$

According to this equation, the drain current I_D is controlled by the gate-source voltage V_{GS} but *independent* of V_{DS}, so we can represent the NMOS with a voltage-controlled current source. This is shown in (a), and details about the parameters $g_m \cdot v_{gs}$ will be presented in the next chapter.

Recall that an ideal current source is a horizontal line on a I_D vs V_{DS} graph, corresponding to an infinite AC resistance. In a real MOSFET, the AC resistance is

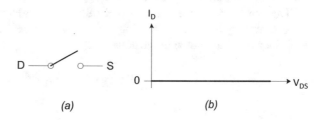

Fig. 6 NMOS cutoff region (**a**) equivalence to open switch (**b**) I-V curve for open switch

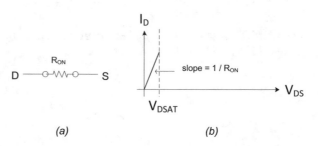

Fig. 7 NMOS triode region (**a**) equivalence to closed switch with finite resistance (**b**) I-V curve for a small on-resistance

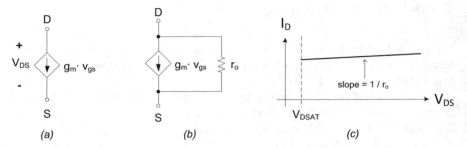

Fig. 8 NMOS saturation region, (**a**) Ideal dependent source (**b**) Improved model (**c**) I-V curve

Fig. 9 NMOS transistor
I-V curves: I_D vs V_{DS}

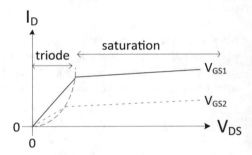

finite and given by r_o in diagram (b). Typically r_o is large, and thus this is represented by a shallow slope in diagram (c) (Fig. 8).

Putting this all together, we have the I-V curve for a NMOS transistor as shown
below. The cutoff region is not shown since it is still a horizontal line at $I_D = 0$. A
family of I-V curves is typically shown in literature, and two I-V curves for different
gate-source voltages are shown where $V_{GS1} > V_{GS2}$ (Fig. 9).

The boundary between triode and saturation is given by V_{DSAT}. As drawn, the
V_{DSAT} boundary is not a vertical line; this is done deliberately since V_{DSAT} gets bigger
as I_D increases, as shown in the equation:

$$I_D = K\left(V_{GS} - V_t\right)^2 = K \cdot V_{DSAT}^2$$

$$V_{DSAT} = \sqrt{\frac{I_D}{K}}$$

For a NMOS in saturation, a higher V_{GS} gives a larger I_D, which causes V_{DSAT} to also
increase as given by the equation above.

5   Designing a Switch with a MOSFET

In this section, we will define a procedure that can be used in designing a MOSFET switch. Specifically, we will design a light-emitting diode (LED) driver using an NMOS switch (Fig. 10).

Step 1 – Review datasheet and note the threshold voltage and on-resistance. Review the LED datasheet and note the forward voltage.

Shown below is a snippet from a NMOS datasheet (Table 3).

Shown below is from a typical blue LED datasheet. Note that the forward voltage drop is much greater than 0.7 V due to the materials used in this device. Different color LEDs will also have different forward voltages, so the designer must look at the relevant LED datasheet (Table 4).

Step 2 – Choose a control input that will turn on the MOSFET.

Using this NMOS, can we drive the gate with 3.3 V logic? (Fig. 11)

Since the NMOS has a minimum V_t of 2 V and a maximum 3.8 V, the 3.3 V gate drive will turn on some of the transistors, but it cannot guarantee that all transistors will be on due to part-to-part variations. We must use logic levels with a voltage higher than the maximum V_t of 3.8 V. In this example, we will choose 5 V logic.

Fig. 10 LED driver circuit using NMOS

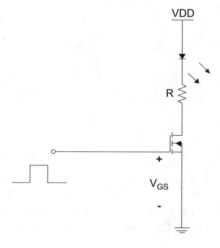

Table 3 NMOS Transistor Datasheet example

Parameter	Test conditions	Min	Max
Gate threshold voltage		2 V	3.8 V
Static drain-source on resistance	$V_{GS} = 10$ V $I_D = 1$ A		0.5 Ω

Table 4 LED Datasheet example

Parameter	Test conditions	Min	Max
Forward voltage	$I_F = 20$ mA	3.2 V	4.0 V

Fig. 11 Controlling the
LED Driver with 3.3 V
logic

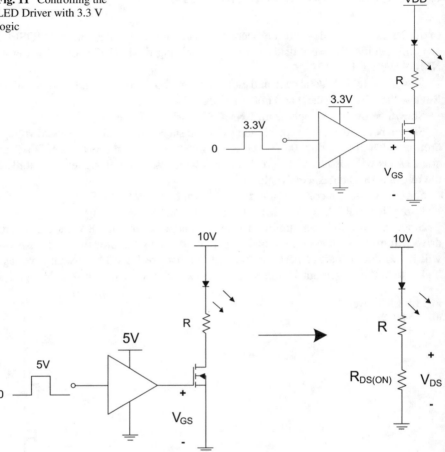

Fig. 12 Analysis of LED driver with 5 V gate drive

Step 3 – Estimate the transistor parameter K from the datasheet. Recall that in triode, the drain-source resistance is estimated by

$$R_{DS(ON)} = \frac{V_{DS}}{I_D} = \frac{1}{2K(V_{GS} - V_t)}$$

Rearranging to solve for K with the datasheet information of $R_{DS(ON)} = 0.5\ \Omega$ max for $V_{GS} = 10$ V and $V_t = 3.8$ V max,

$$K = \frac{1}{2R_{DS(ON)}(V_{GS} - V_t)} = \frac{1}{2(0.5\,\Omega)(10 - 3.8\,\text{V})} = 161\frac{\text{mA}}{\text{V}^2}$$

Step 4 – Calculate $R_{DS(ON)}$.

Using 5 V logic for gate drive, $V_{GS} = 5$ V, and thus we can recalculate $R_{DS(ON)}$

$$R_{DS(ON)} = \frac{1}{2(0.161)(5-3.8)} = 2.6\,\Omega$$

Step 5 – Verify that the NMOS works as a switch for the given load specification, where $V_{DD} = 10$ V, and we desire 5 mA through the LED (Fig. 12).

We can replace the NMOS with its equivalent on-resistance to perform the analysis. Given that the maximum LED voltage drop is 4 V,

$$R_{DS(ON)} + R = \frac{10-4\,V}{5\,mA} = 1.2\,k\Omega$$

Since $R_{DS(ON)}$ is much smaller than 1.2 kΩ, we can choose $R = 1.2$ kΩ. The resulting $V_{DS} = 10$ V$\cdot$2.6 Ω/(1.2 kΩ + 2.6 Ω) = 21.6 mV. Since $V_{DSAT} = 5$ V $-$ 3.8 V $= 1.2$ V, the NMOS is in triode as desired.

6 Problems

1. What is the DC current into the gate of an NMOS transistor?
2. When measuring the impedance between the gate and source of an MOS transistor, which would we expect to see (resistive, inductive, or capacitive)?
3. In the four-terminal symbol for a MOSFET, what is the body or bulk terminal connected to?
4. In a discrete MOS transistor, there are only three terminals: drain, gate, and source. What happened to the body or bulk terminal?
5. An NMOS transistor is in the common source configuration:

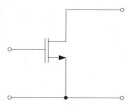

Determine the region of operation (cutoff, triode, saturation) for each of the following conditions. Assume that threshold voltage is $V_t = 1.5$ V:

(a) $V_{GS} = 3$ V, $V_{DS} = 2$ V.
(b) $V_{GS} = 0$ V, $V_{DS} = 2$ V.

(c) $V_{GS} = 5$ V, $V_{DS} = 2$ V.

(d) $V_{GS} = 0.5$ V, $V_{DS} = 5$ V.

(e) $V_{GS} = 5$ V, $V_{DS} = 5$ V.

6. For each circuit below, determine the DC operating point (I_D, V_{DS}) and the region of operation. Assume that $V_t = 2.5$ V and $K = 25$ mA/V^2.

(a)

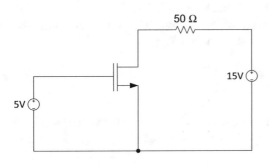

(b)

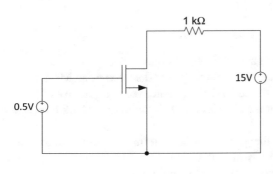

(c)

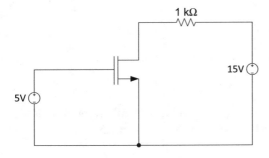

7. An NMOS transistor is specified in the datasheet as having a threshold voltage $V_t = 2$ V and $I_D = 10$ mA for $V_{GS} = 4$ V in the saturation region. Find the parameter K from this information.

8. The I-V curve of a NMOS transistor is shown below. From the information, find the AC resistance in the triode region and saturation region. Estimate the value of λ from the data and your calculations.

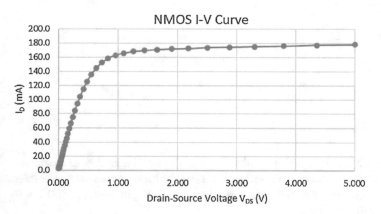

VDS(V)	ID (mA)	VDS(V)	ID (mA)	VDS(V)	ID (mA)
0.005	2.4	0.32	93.8	1.7	170.7
0.010	4.2	0.42	114.9	1.9	171.8
0.020	7.6	0.55	135.5	2.2	172.5
0.030	11.1	0.72	152.4	2.5	173.4
0.040	14.4	0.83	158.4	2.9	174.3
0.060	21.2	0.95	162.9	3.3	175.1
0.079	27.4	1.09	165.7	3.8	176.0
0.104	35.5	1.26	168.0	4.4	177.1
0.208	66.5	1.44	169.4	5.0	178.0

9. We wish to design a switch that interfaces a 5 V logic signal to a load connected to +15 V. Assuming that V_{DS} is small, use the approximate method to find the on-resistance and the DC operating point of the transistor when the input is +5 V. The NMOS transistor has $K = 8$ mA/V^2 and $V_t = 1.2$ V.

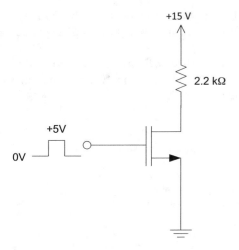

10. Tinkercad®: Measure the threshold voltage of the small signal NMOS transistor using the circuit below. We will define V_t as the highest gate-source voltage that has a drain current of zero; test your circuit with V_{GS} increments of 0.1 V to find the threshold voltage.

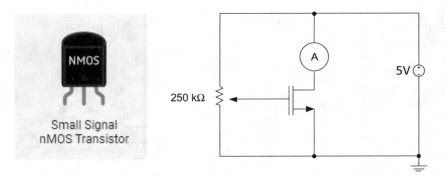

11. Tinkercad®: Measure I_D for multiple values of V_{GS}. Using this and the threshold voltage found in the previous question, estimate the value of K for this NMOS transistor.

12. Tinkercad®: Simulate the circuit below using the small signal NMOS transistor, and record I_D and V_{DS}. Compare this result against prediction using K and V_t obtained in previous questions.

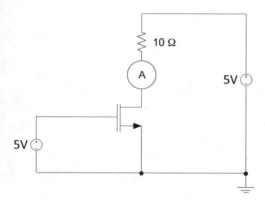

13. Tinkercad®: Simulate the circuit below using the small signal NMOS transistor, and record I_D and V_{DS}. Compare this result against prediction using K and V_t obtained in previous questions.

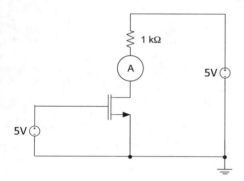

14. Tinkercad®: Measure the forward voltage V_F of a LED as shown in the test circuit.

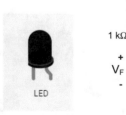 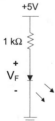

15. Tinkercad®: Design an LED driver using an NMOS transistor. The LED runs off a +5 V power supply with a target of 5 mA when on. The NMOS transistor is controlled using a 3.3 V logic signal

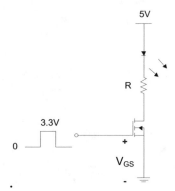

Single-Transistor Amplifiers

In this chapter, we begin the study of MOSFET amplifiers. We will discuss small signal models, which are linear representations of an intrinsically nonlinear transistor. Having a linear model of a transistor greatly simplifies the circuit analysis. We start by focusing on amplifiers that only use a single transistor, of which there are three circuits:

- Common source amplifier
- Common gate amplifier
- Common drain amplifier (aka the Source Follower)

1 Voltage-Mode vs Current-Mode Amplifiers

Before diving in, we would like to gain some appreciation of what characteristics an *ideal* amplifier should have. Moreover, our signal of interest may be in the form of a voltage or a current. Thus, there are amplifiers that amplify voltages, and there are some that amplify currents; the ideal characteristics for each type are quite different from each other.

A linear voltage-mode amplifier accepts an AC input voltage and outputs an AC voltage. As shown in the figure below, the amplifier has a voltage gain A_V, and if the input has a peak of V_{PK}, the output peak is $A_V \cdot V_{PK}$ (Fig. 1).

A linear current-mode amplifier accepts an AC input current and outputs an AC current. As shown in the figure below, the amplifier has a current gain A_I, and if the input has a peak of I_{PK}, the output peak is $A_I \cdot I_{PK}$ (Fig. 2).

To understand what characteristics each type of amplifier should have, we will create a simple model and consider the effect of *amplifier loading*.

© Springer Nature Switzerland AG 2022
C. Siu, *Electronic Devices, Circuits, and Applications*,
https://doi.org/10.1007/978-3-030-80538-8_6

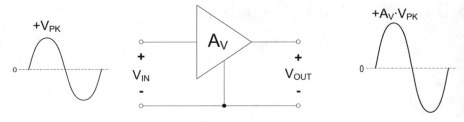

Fig. 1 Voltage mode amplifier

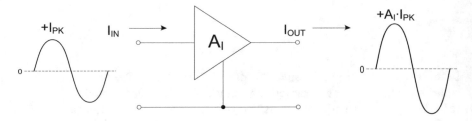

Fig. 2 Current mode amplifier

Voltage-Mode Amplifier

We can model a voltage amplifier with gain A_V as follows:

- r_{in} is the input resistance of the amplifier.
- r_{out} is the output resistance of the amplifier.
- The voltage-controlled voltage source $A_V \cdot v_{in}$ models the voltage gain (Fig. 3).

Loaded Gain of Voltage Amplifiers

In actual applications, the amplifier drives a load R_L while connected to a signal source with resistance R_S. What is the voltage gain in this case, also referred to as the *loaded gain*? (Fig. 4)

To find an expression for the loaded gain V_{OUT}/V_S, we can replace the amplifier with its model and analyze the equivalent circuit (Fig. 5):

There is voltage division between the signal source R_S and the amplifier input resistance

$$V_{IN} = \frac{r_{in}}{R_S + r_{in}} \cdot V_S$$

On the output side, the dependent source's output is divided between the amplifier output resistance and the load resistance

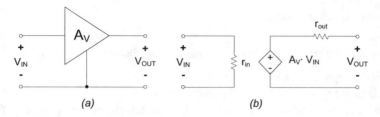

Fig. 3 Voltage mode amplifier (**a**) symbol (**b**) model

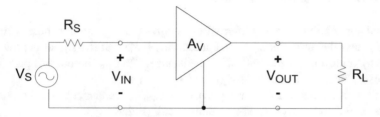

Fig. 4 Voltage mode amplifier with source & load

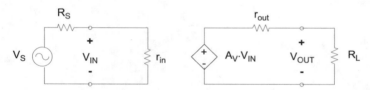

Fig. 5 Voltage mode amplifier analysis

$$V_{OUT} = \frac{R_L}{R_L + r_{out}} \cdot A_V V_{IN}$$

Combining the two equations, the expression for loaded voltage gain is

$$A_{V(LOADED)} = \frac{V_{OUT}}{V_S} = \frac{r_{in}}{R_S + r_{in}} \cdot \frac{R_L}{R_L + r_{out}} \cdot A_V$$

We can see that amplifier loading reduces the gain, as $A_{V(LOADED)} \leq A_V$. The loaded gain is maximized and approaches A_V if the following conditions are met:

1. The amplifier input resistance is zero $r_{in} \rightarrow \infty$.
2. The amplifier output resistance is infinite $r_{out} = 0$.

.*Current-Mode Amplifier*

We can model a current amplifier with gain A_I as follows:

- r_{in} is the input resistance of the amplifier.
- r_{out} is the output resistance of the amplifier.
- The current-controlled current source $A_I \cdot i_{in}$ models the current gain (Fig. 6).

Loaded Gain of Current Amplifiers

In actual applications, the amplifier will be connected to a signal source with resistance R_S, and the amplifier will drive a load R_L. Note that the signal is now represented by a current source with finite resistance. What is the current gain in this case, also referred to as the *loaded gain*? (Fig. 7)

To find an expression for the loaded gain I_{OUT}/I_S, we can replace the amplifier with its model and analyze the equivalent circuit (Fig. 8):

There is current division between the signal source R_S and the amplifier input resistance

$$I_{IN} = \frac{R_S}{R_S + r_{in}} \cdot I_S$$

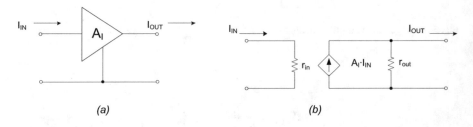

(a)   *(b)*

Fig. 6 Current mode amplifier (**a**) symbol (**b**) model

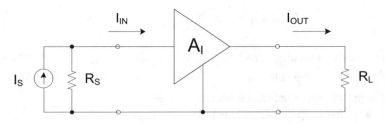

Fig. 7 Current mode amplifier with source & load

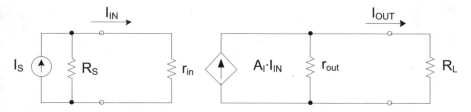

Fig. 8 Current mode amplifier analysis

On the output side, the dependent source's output is divided between the amplifier output resistance and the load resistance.

$$I_{OUT} = \frac{r_{out}}{R_L + r_{out}} \cdot A_I I_{IN}$$

Combining the two equations, the expression for loaded current gain is

$$A_{I(LOADED)} = \frac{I_{OUT}}{I_S} = \frac{R_S}{R_S + r_{in}} \cdot \frac{r_{out}}{R_L + r_{out}} \cdot A_I$$

We can see that amplifier loading reduces the gain, as $A_{I(LOADED)} \leq A_I$. The loaded gain is maximized and approaches A_I if the following conditions are met:

1. The amplifier input resistance is zero $r_{in} = 0$.
2. The amplifier output resistance is infinite $r_{out} \to \infty$.

2   DC Biasing for Amplifier Design

An important step in amplifier design is to choose the DC operating point of the transistor. Obviously, the MOSFET must operate in the *saturation* region, but what value of I_D and V_{GS} should be used? It turns out that the DC operating point has a direct influence on amplifier performance, and so in this section we will study circuit topologies that can be used to reliably establish a DC operating point.

Drain Feedback Resistor

The first DC bias circuit uses a large resistance connected between gate and drain (Fig. 9).

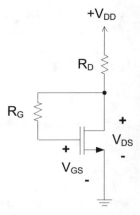

Fig. 9 DC bias circuit with drain feedback resistor

Since the DC gate current into the NMOS is zero, there is no voltage drop across R_G. As a result, $V_{DS} = V_{GS}$, and since $V_{DSAT} = V_{GS} - V_t$, then combining the two expressions we have $V_{DS} = V_{DSAT} + V_t$. Hence, $V_{DS} > V_{DSAT}$, and this circuit guarantees that the NMOS is in saturation by design.

Exercise Find the DC operating point for the NMOS below. The transistor parameters are as follows:

- $K = 0.24$ mA/V^2
- $V_t = 3$ V

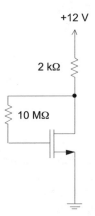

Answer From the main text, we know that the NMOS is in saturation; hence the saturation equation is used.

$$I_D = K\left(V_{GS} - V_t\right)^2$$

There are two unknowns in this equation: I_D and V_{GS}. A second equation can be obtained by recalling that for this circuit, $V_{GS} = V_{DS}$. Hence, the voltage drop across the 2 kΩ resistor is 12 V − V_{GS}. The current through this resistor is equal to I_D, hence by Ohm's law

$$I_D = \frac{12\,\text{V} - V_{GS}}{2\,k\Omega}$$

The two expressions can be equated to each other, and substituting known quantities, we arrive at a quadratic equation in V_{GS}.

$$\frac{12\,\text{V} - V_{GS}}{2\,k\Omega} = K\left(V_{GS} - V_t\right)^2$$

$$\frac{12\,\text{V} - V_{GS}}{2\,k\Omega} = \left(0.24\,\frac{\text{mA}}{\text{V}^2}\right)\left(V_{GS} - 3\,\text{V}\right)^2$$

$$0.48 \cdot V_{GS}^2 - 1.88 \cdot V_{GS} - 7.68 = 0$$

The two solutions are V_{GS} = +6.4 V and −2.5 V. The negative solution is invalid since that would mean the NMOS is off. Hence, $V_{GS} = V_{DS}$ = 6.4 V is the valid solution, and substituting this back into an earlier equation yields I_D = 2.8 mA. The NMOS DCOP is 2.8 mA, 6.4 V.

Voltage Divider Bias

Although the drain feedback resistor circuit guarantees that the NMOS will be in saturation, it does constrain the drain-to-source voltage to be equal to V_{GS}. This limits the amplifier's output voltage swing, and so here is another DC bias circuit that does not have this constraint. Note that a voltage divider sets the bias voltage on the gate terminal, and a resistor R_S is in series with the source terminal. Without proof, it will just be stated that R_S is used to make the DC operating point less sensitive to transistor parameters (e.g., K, V_t), which makes the amplifier more consistent during mass production (Fig. 10).

We will illustrate the analysis of this bias circuit using an example.

Fig. 10 DC bias circuit
with gate voltage divider

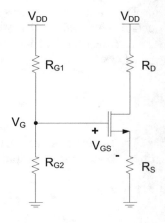

Exercise Find the DC operating point for the NMOS below. The transistor parameters are as follows:

- $K = 0.12$ mA/V^2
- $V_t = 5$ V

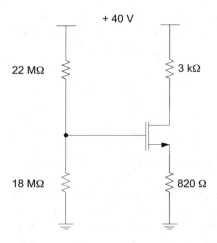

Answer Since the DC gate current is zero, the gate voltage can be calculated using
voltage division, giving $V_G = 18$ V. Using KVL, the voltage across the 820 Ω resistor
can be expressed as 18 V $- V_{GS}$.

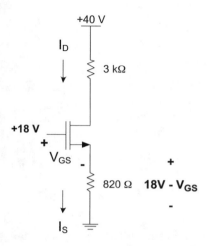

We can now use Ohm's law to express the current through the 820 Ω resistor, which is I_S. Before doing so, we will take an extra step and note that $I_S = I_D$ since $I_G = 0$ at DC. Putting all this together yields

$$I_D = \frac{18\,\text{V} - V_{GS}}{820\,\Omega}$$

There are two unknowns I_D and V_{GS}, and so we need a second equation. This is set by our design objective: for use as an amplifier, we need the NMOS in saturation

$$I_D = K\left(V_{GS} - V_t\right)^2$$

Equating the two expressions and simplifying, we will obtain a quadratic equation in V_{GS}. From this, the NMOS DCOP is found to be 6.71 mA, 14.4 V.

Gate Pulldown Bias

If a dual polarity supply is available, it is possible to bias an NMOS in saturation with the following circuit (Fig. 11):

The NMOS gate terminal is biased at 0 V via a large resistance to ground. For a positive V_{GS}, this means that the NMOS source voltage V_S is negative, made possible by the use of a negative power supply $-V_{SS}$. The current source I_{DC} is typically implemented with transistors, though for now we will just assume that it is an ideal current source.

Fig. 11 DC bias circuit
with gate pulldown resistor

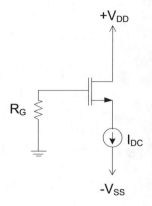

Exercise Find the DC operating point for the NMOS below. The transistor parameters are as follows:

- $K = 0.5$ mA/V^2
- $V_t = 1.5$ V

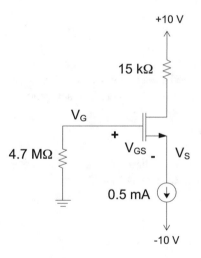

Answer Since the gate current $I_G = 0$, there is no voltage drop across the 4.7 MΩ resistor, and $V_G = 0$ V. The drain current I_D is set by the current source, which is 0.5 mA. We will calculate V_{GS} by assuming that the NMOS is in saturation and then checking this assumption later.

$$I_D = K\left(V_{GS} - V_t\right)^2$$

$$0.5\,\text{mA} = \left(0.5\frac{\text{mA}}{\text{V}^2}\right)\left(V_{GS} - 1.5\,\text{V}\right)^2$$

$$V_{GS} = 2.5\,\text{V}$$

We can now find the source voltage $V_S = V_G - V_{GS} = 0\,\text{V} - 2.5\,\text{V} = -2.5\,\text{V}$.

The drain voltage V_D is found by $10\,\text{V} - (0.5\,\text{mA})(15\,\text{k}\Omega) = +2.5\,\text{V}$.

The drain-source voltage V_{DS} is thus $2.5\,\text{V} - (-2.5\,\text{V}) = 5\,\text{V}$. With $V_{DSAT} = 2.5\,\text{V} - 1.5\,\text{V} = 1\,\text{V}$, the NMOS is in saturation, and our assumption is correct. The NMOS DCOP is 0.5 mA, 5 V.

3 NMOS Small Signal Model

After establishing a DC operating point in the saturation region, we can use the MOSFET as an amplifier. For example, we start with the voltage divider bias circuit inside the rectangle below. Then the input signal is AC coupled into the gate terminal, and the output signal is AC coupled from the drain terminal, forming what is known as a *common-source* amplifier (Fig. 12).

Note that there is a capacitor connected between the NMOS source terminal and ground. The capacitor is assumed to be big enough such that it can be approximated as an AC short circuit, creating an AC ground at the source terminal. Looking at the common-source amplifier more closely, we have the following situation:

- We will apply a sine wave to the amplifier input; the sine wave magnitude is assumed to be tiny enough such that the **small signal** approximation is valid (we will explain what that means a bit later).

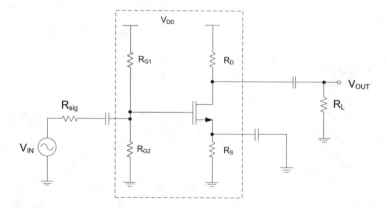

Fig. 12 Voltage divider bias circuit configured into an amplifier

- The sine wave causes the gate-to-source voltage v_{GS} to change with time, which in turn makes the drain current i_D vary with time..., and hopefully this creates amplification.
- We will look for a compact way of expressing how i_D changes with v_{GS}, under the small signal assumption.

Transconductance of a MOSFET

Since the NMOS is being used as an amplifier, the transistor must have some intrinsic gain. *Transconductance* is one of the ways that transistor gain is specified, with **g_m** being a common symbol for transconductance. Transconductance measures how much the *output current* changes for a given *input voltage* change, and hence g_m has units of Ampere/Volt.

Applying this idea to a NMOS in saturation, if we treat the input as V_{GS} and the output as I_D, then we already have an equation relating the two quantities:

$$I_D = K\left(V_{GS} - V_t\right)^2$$

The equation is plotted below as depicted by the thick solid line (note that this is a graph of I_D vs V_{GS}, and not I_D vs V_{DS} as shown in the earlier I-V curve section) (Fig. 13).

Suppose that we have biased the NMOS at some DC operating point with a drain current of I_{D1}, marked with a dot on the I_D-V_{GS} curve. If we vary the input V_{GS}, how much does the output I_D change? The **small signal approximation** narrows this question further by constraining the V_{GS} change to be small. The answer is obtained by calculus, by finding the derivative or slope to that point on the curve. Hence, differentiating the saturation equation will give us a formula for g_m.

$$g_m = \frac{dI_D}{dV_{GS}} = 2K\left(V_{GS} - V_t\right)$$

Fig. 13 Transconductance of a NMOS transistor

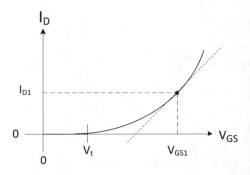

Low-Frequency, Small Signal Model of NMOS Transistor

With transconductance defined under the small signal approximation, we can now create a *small signal model* for the NMOS (Fig. 14).

Since the DC gate current is zero, the gate terminal is an open circuit.

For a small change in V_{gs}, the drain current I_d will vary in accordance to g_m; this is represented by the voltage-controlled current source between drain and source.

A real transistor deviates from an ideal current source in that the AC resistance between drain to source is not infinite. This is modeled by the resistor r_o. Without proof, the equation for r_o is given by

$$r_o = \frac{1}{\lambda \cdot I_D}$$

where λ is a transistor parameter that is provided to us. We will discuss λ in greater detail in a later chapter.

Exercise For an NMOS with $K = 20$ mA/V^2, $V_t = 1.5$ V, and $\lambda = 0.1$, find the trans-conductance and drain AC resistance if $I_D = 0.5$ mA with the transistor in saturation.

Answer For $I_D = 0.5$ mA, we can first solve for V_{GS}

$$V_{GS} = \sqrt{\frac{I_D}{K}} + V_t = \sqrt{\frac{0.5\,\text{mA}}{20\,\text{mA}}} + 1.5\,\text{V} = 1.66\,\text{V}$$

The transconductance $g_m = 2(20\text{ m})(1.66–1.5) = 6.4$ mS. The drain AC resistance $r_o = 1/(0.1 \cdot 0.5$ mA$) = 20$ kΩ.

Fig. 14 NMOS low frequency, small signal model

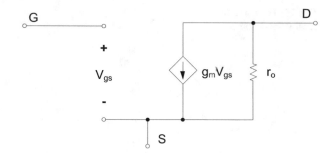

4 AC Ground and AC Resistances

We are about to begin the analysis of MOSFET amplifiers. Before starting, the ideas of AC resistance and capacitive reactance will be reviewed.

For an ideal voltage source, the voltage does not change regardless of how much the current varies. Hence, the AC resistance of an ideal voltage source is *zero* (Fig. 15).

For an ideal current source, the current does not change regardless of how much the voltage varies. Hence, the AC resistance of an ideal current source is *infinite* (Fig. 16).

Next, recall that the reactance of a capacitor is given by

$$X_c = \frac{1}{\omega C} = \frac{1}{2\pi fC}$$

If the signal frequency f and the capacitance C are large, the reactance X_C is very small, almost like a short circuit for the AC signal. We will call this an *AC short circuit* or *AC short*.

Furthermore, if one terminal of the capacitor is connected to ground, then the capacitor can create an AC short circuit to ground, which we will call *AC ground* in subsequent work (Fig. 17).

In amplifier analysis, we are often interested in the circuit response to an AC signal, for example, the amplifier gain for a 1 kHz sine wave input. Recall that an ideal DC current source has an infinite AC resistance; it looks like an open circuit to an AC signal. Is it OK to remove the current source for AC gain analysis?

The answer is yes, and it leads us to the technique of AC equivalent circuits. The idea is to remove or replace components in the schematic that has a known effect on AC signals, in the process simplifying the analysis. The example below will illustrate the procedure.

Fig. 15 I-V curve for an ideal voltage source

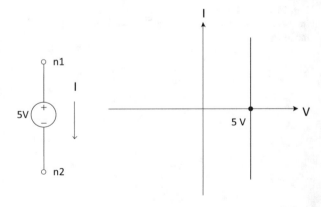

Fig. 16 I-V curve for an
ideal current source

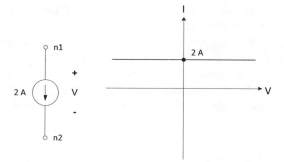

Fig. 17 Concept of an AC
ground

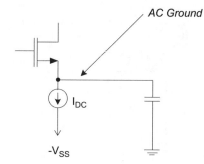

Exercise Draw the AC equivalent circuit for the schematic below. Assume that the
capacitors are large enough such that at the frequencies of interest, they are AC short
circuits.

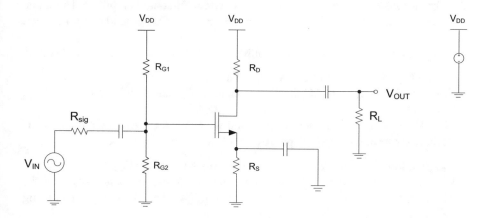

Answer The three capacitors shown can be replaced by wires in the AC equivalent
circuit. Note that the capacitor connected to the NMOS source terminal is going to
ground, and hence there is an *AC ground* on the NMOS source.

In addition, the V_{DD} connections at the top of the schematic are also replaced with AC grounds. Since V_{DD} is connected to an ideal voltage source (DC power supply), there is zero AC resistance to ground. This is emphasized here with the V_{DD} voltage source on the schematic, but it is usually omitted with the understanding that there is a DC power supply connected to V_{DD}.

The AC equivalent circuit is shown below. Note that with the AC ground on the NMOS source, R_S is shorted out and can be deleted.

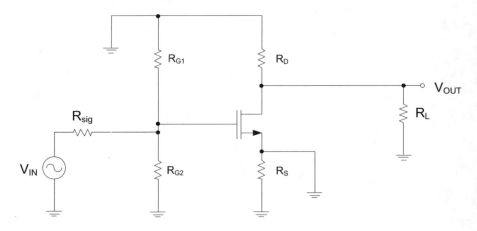

5   Common Source Amplifier

The first MOSFET amplifier topology is the common-source amplifier. The name comes from the fact that the NMOS source terminal is at AC ground or *AC common* (Fig. 18).

We perform the analysis of this amplifier under the small signal approximation:

- Replace the NMOS with its small signal model.
- Identify the AC ground, AC short-circuits, and AC open-circuits in the amplifier, and complete the AC equivalent circuit diagram.
- The AC equivalent is a linear circuit, and we can use the usual circuit analysis techniques to find the gain, input resistance, and output resistance.

AC Equivalent Circuit for the Common Source Amplifier (Fig. 19)

Take a moment and check that you understand the correspondence between the original schematic and the AC equivalent circuit. The analysis will proceed using the equivalent circuit.

First on the input side, note that there is voltage division between R_{sig} and the parallel combination of R_{G1} and R_{G2}

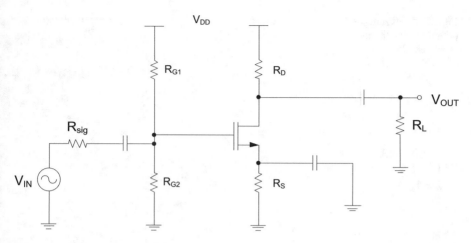

Fig. 18 Common source amplifier

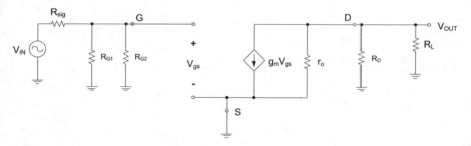

Fig. 19 AC equivalent circuit for the common source amplifier

$$v_{gs} = \frac{R_{G1} \parallel R_{G2}}{R_{sig} + R_{G1} \parallel R_{G2}} \cdot v_{in}$$

Next on the output side, the current from the dependent source flows through the parallel combination of r_o, R_D, and R_L. Note that with the polarity of V_{out} defined, the direction of current flow creates a negative sign in the equation (Fig. 20):

$$v_{out} = -g_m v_{gs} \left(r_o \parallel R_D \parallel R_L \right)$$

Combining the two equations, we obtain an expression for the small-signal voltage gain A_V of the common source amplifier

$$A_v = \frac{v_{out}}{v_{in}} = -\frac{R_{G1} \parallel R_{G2}}{R_{sig} + R_{G1} \parallel R_{G2}} \cdot g_m \left(r_o \parallel R_D \parallel R_L \right)$$

Exercise   Find the transconductance g_m and the AC drain resistance r_o of the NMOS below. The transistor parameters are as follows:

Fig. 20 Analyzing the output of the common source amplifier

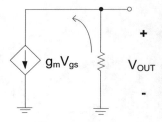

- $K = 0.5\ \text{mA/V}^2$
- $V_t = 1.5\ \text{V}$
- $\lambda = 1/75$

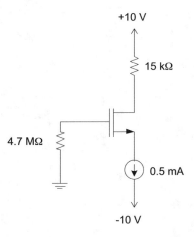

Answer From an earlier example in the gate pulldown bias section, we found the DC operating point of the NMOS to be 0.5 mA, 5 V with $V_{\text{DSAT}} = 1$ V.

$$g_m = 2\left(0.5\frac{\text{mA}}{\text{V}^2}\right)(2.5\,\text{V} - 1.5\,\text{V}) = 1\,\text{mS}$$

$$r_o = \frac{1}{\lambda \cdot I_D} = \frac{75}{0.5\,\text{mA}} = 150\,k\Omega$$

Exercise The bias circuit in the previous example is configured to be a common-source amplifier. Find the small signal voltage gain A_V for the unloaded case and for the loaded case with $R_{\text{sig}} = 100$ kΩ and $R_L = 15$ kΩ.

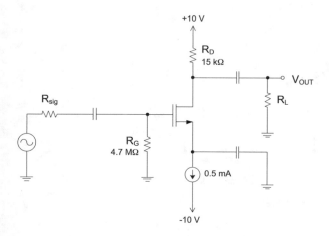

Answer   Using the g_m and r_o values from the previous example, we can plug the numbers into the following equation for A_V:

$$A_v = \frac{v_{out}}{v_{in}} = -\frac{R_G}{R_{sig} + R_G} \cdot g_m \left(r_o \parallel R_D \parallel R_L \right)$$

Unloaded Gain

For the first case, $R_{sig} = 0$ and $R_L \to \infty$. Substituting,

$$A_v = -\frac{4.7\,M}{0 + 4.7\,M} \cdot (1\,m)(150k \parallel 15k \parallel \infty) = -13.6\,\frac{V}{V}$$

Loaded Gain

With $R_{sig} = 100\ k\Omega$ and $R_L = 15\ k\Omega$, $A_V = -7$ V/V. Note that with loading, the magnitude of the amplifier gain has decreased. Relate this to the ideas introduced at the beginning of the chapter, with regard to amplifier loading.

CS Amplifier Input and Output Resistance

Recall that an ideal voltage amplifier has an input resistance $r_{in} \to \infty$ and $r_{out} = 0$. If this can be achieved, then the amplifier gain will not be degraded by loading. In the previous example, we saw that gain was degraded by loading; hence the common-source amplifier does not have the ideal r_{in} and r_{out} (Fig. 21).

In the schematic above, a common-source amplifier is shown within the dotted lines. If we measure the AC resistance at the amplifier input, we would obtain r_{in}; note that r_{in} does not include the signal generator resistance R_{sig}. If we measure the

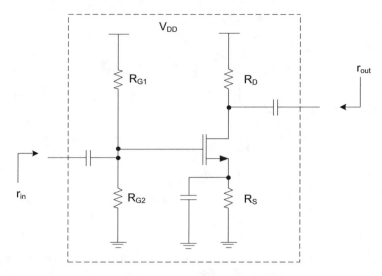

Fig. 21 Input & output resistances of a common source amplifier

AC resistance at the amplifier output, we would obtain r_{out}, noting that r_{out} does not include the load resistance R_L.

One way to find r_{in} and r_{out} is to use the AC equivalent circuit. The equivalent circuit for the common-source amplifier is shown below, but without the signal source and the load (Fig. 22).

At the input, we can see that $r_{in} = R_{G1} \parallel R_{G2}$ by inspection.

At the output, we see r_o in parallel with R_D, which is also in parallel with the dependent source. Recall that an ideal current source, even dependent sources, has an infinite AC resistance. Hence, $r_{out} = r_o \parallel R_D$.

Finally, let us look at the gain equation again and rewrite it in terms of r_{in}:

$$A_v = -\frac{R_{G1} \parallel R_{G2}}{R_{sig} + R_{G1} \parallel R_{G2}} \cdot g_m \left(r_o \parallel R_D \parallel R_L \right) = -\frac{r_{in}}{R_{sig} + r_{in}} \cdot g_m \left(r_o \parallel R_D \parallel R_L \right)$$

Note that the first term in the equation quantifies the signal loss due to voltage division between the signal generator R_{sig} and the amplifier input r_{in}. This will be a recurring theme as we analyze the other amplifier topologies.

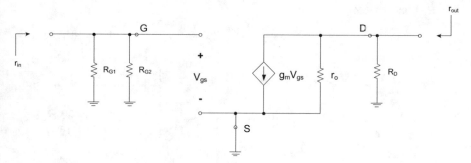

Fig. 22 Analysis of r_{in} and r_{out} for a common source amplifier

Exercise What is r_{in} and r_{out} for the amplifier below?

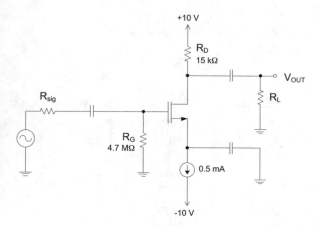

Answer Please refer to the previous example for some relevant values. In this circuit, $r_{in} = R_G = 4.7$ MΩ, and $r_{out} = r_o \parallel R_D = 150$ k$\Omega \parallel 15$ k$\Omega = 13.6$ kΩ.

6 Common Gate Amplifier

The next topology we will study is the common-gate amplifier. Compared to the common-source amplifier, notice that two things have changed: the signal input is now connected to the NMOS source terminal, while the capacitor connected to the gate creates an AC ground at that node. The name of this topology thus comes from the NMOS gate terminal being at AC common (Fig. 23).

To create the AC equivalent circuit, we follow the same steps as before:

- All DC power supply connections become AC ground.
- Capacitors are assumed to the AC short circuits.

Fig. 23 Common gate amplifier

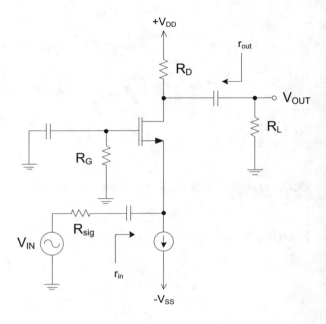

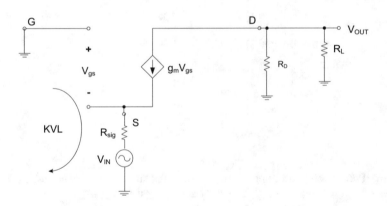

Fig. 24 AC equivalent circuit for the common gate amplifier

- Replace the NMOS with its small signal model; however for this analysis, we will assume $r_o \to \infty$, since its inclusion creates a fog of math that is not suitable for the introduction here.

AC Equivalent Circuit for the Common Gate Amplifier (Fig. 24)

The analysis begins with KVL on the input side as shown, noting that the current through R_{sig} is equal to the dependent source $g_m \cdot v_{gs}$:

$$v_{gs} + g_m v_{gs} R_{sig} + v_{in} = 0$$

From this, we create an expression for v_{gs}:

$$v_{gs} = \frac{-v_{in}}{1 + g_m R_{sig}}$$

On the output side, the dependent source current goes through $R_D \parallel R_L$ to generate v_{out}:

$$v_{out} = -g_m v_{gs} \left(R_D \parallel R_L \right)$$

Combining the two equations provides us with the gain for the common-gate amplifier:

$$A_v = \frac{v_{out}}{v_{in}} = + \frac{1}{1 + g_m R_{sig}} \cdot g_m \left(R_D \parallel R_L \right)$$

If we compare this to the common-source amplifier gain, one of the differences is the sign; the common-gate amplifier has a positive gain, whereas the common-source amplifier has a negative gain. This sign indicates the phase relationship between the input and output. If the gain has a negative sign, then the output is 180° out of the phase relative to the input. If the gain has a positive sign, then the output and input are in-phase with respect to each other. The figure below illustrates this with a gain of +10 and −10 amplifiers (Fig. 25).

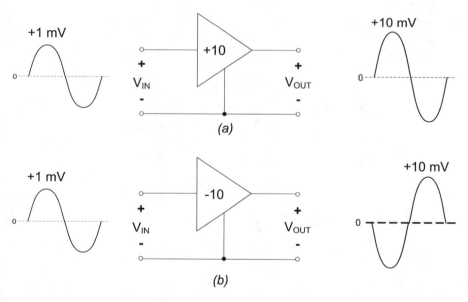

Fig. 25 Input and output for (a) non-inverting amplifier (b) inverting amplifier

CG Amplifier Input and Output Resistance

The input signal is now connected to the source terminal of the NMOS. What is the AC resistance presented by the NMOS source terminal, as this determines the amplifier input resistance? We may be tempted to reach for an intuitive answer by inspection, such as r_o, but intuition would be wrong in this case. Circuit analysis is required to arrive at the correct answer, and this is presented in a later chapter. For now, the result is presented without proof: the NMOS source terminal has an AC resistance of $1/g_m$, assuming that the device is in saturation. Hence, the input and output resistances of the common-gate amplifier are

$$r_{in} = \frac{1}{g_m}$$

$$r_{out} = R_D$$

Exercise Find input resistance and output resistance of the common-gate amplifier shown. Then, calculate the voltage gain A_V for the unloaded case, and for the loaded case with $R_{sig} = 100 \text{ k}\Omega$ and $R_L = 15 \text{ k}\Omega$

- $K = 0.5 \text{ mA/V}^2$
- $V_t = 1.5 \text{ V}$
- $\lambda = 0$

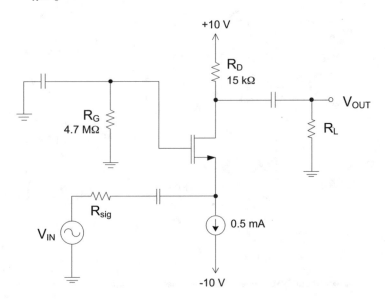

Answer The DC bias circuit is the same as the example in the common-source amplifier section so we can reuse those results, summarized here for convenience:

- DCOP = (0.5 mA, 5 V)
- $V_{DSAT} = 1$ V
- $g_m = 1$ mS

The input resistance $r_{in} = 1/(1$ mS$) = 1$ kΩ, and the output resistance $r_{out} = 15$ kΩ.

Unloaded Gain

For the first case, $R_{sig} = 0$ and $R_L \rightarrow \infty$. Substituting,

$$A_v = +\frac{1}{1 + g_m R_{sig}} \cdot g_m \left(R_D \parallel R_L \right) = \frac{1}{1+0}(1m)(15k) = +15 \text{ V/V}$$

Loaded Gain

$$A_v = +\frac{1}{1 + 1m \cdot 100k} \cdot (1m)(15k \parallel 15k) = +0.074 \text{ V/V}$$

In the unloaded gain, we have a healthy gain, but once loading is added, the voltage gain plummets to the point where the circuit is attenuating the input signal. Most of the gain loss is due to the low rin of the amplifier, which causes severe voltage division with $R_{sig} = 100$ kΩ. Is there a useful application for the common-gate amplifier? We will delve into this at the end of the chapter.

Before leaving the common-gate amplifier, let us look at the gain equation again and write it in a slightly different way.

$$A_v = +\frac{1}{1 + g_m R_{sig}} \cdot g_m \left(R_D \parallel R_L \right) = \frac{1/g_m}{1/g_m + R_{sig}} \cdot g_m \left(R_D \parallel R_L \right)$$

Since $r_{in} = 1/g_m$,

$$A_v = \frac{r_{in}}{r_{in} + R_{sig}} \cdot g_m \left(R_D \parallel R_L \right)$$

The first term in the gain equation represents the voltage division between the signal generator R_{sig} and the amplifier input r_{in}. This is similar to what we saw before in the common-source amplifier equation.

7   Common Drain Amplifier

The third amplifier topology has the drain terminal at AC ground. This circuit is also known as a source follower, for reasons that will be clear after completing the analysis (Fig. 26).

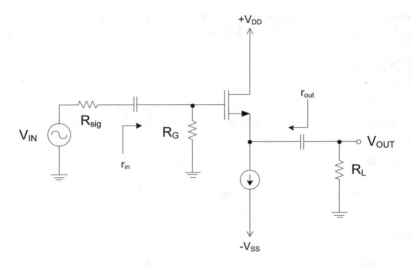

Fig. 26 Common drain amplifier or source follower

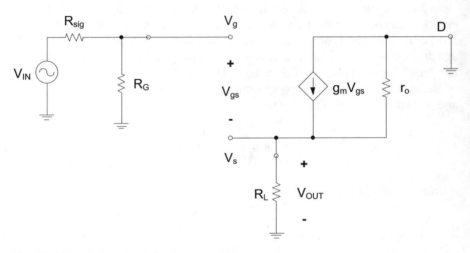

Fig. 27 AC equivalent circuit for the source follower

AC Equivalent Circuit for the Common Drain Amplifier (Fig. 27)

On the input side, the signal generator V_{in} goes through a voltage divider to create v_g (not v_{gs}, since the source terminal is not at AC ground):

$$v_g = \frac{R_G}{R_{sig} + R_G} v_{in}$$

Next by inspection or KVL,

$$v_g = v_{gs} + v_{out}$$

Equating these two expressions gives

$$\frac{R_G}{R_{sig} + R_G} v_{in} = v_{gs} + v_{out}$$

At the output, notice how R_L and r_o are actually in parallel, and current from the dependent source flows through the resistors to yield v_{out}

$$v_{out} = +g_m v_{gs} \left(r_o \| R_L \right)$$

These equations can be combined to eliminate v_{gs} and create an expression for the voltage gain

$$A_v = \frac{v_{out}}{v_{in}} = +\frac{R_G}{R_{sig} + R_G} \cdot \frac{r_o \| R_L}{\frac{1}{g_m} + r_o \| R_L}$$

Notice that the gain equation consists of two voltage divider ratios, meaning that the maximum value of A_v is +1. This is the desired gain for a common-drain amplifier, having the output follow the input exactly with a gain of 1. The more common name of this topology is thus the *source follower*.

Source Follower Input and Output Resistance

Recall that the MOSFET source terminal has a small signal resistance of $1/g_m$, and this is equal to the output resistance r_{out}. The input resistance r_{in} can be obtained by inspecting the equivalent circuit and is equal to the large pulldown resistor R_G. Thus,

$$r_{in} = R_G$$

$$r_{out} = \frac{1}{g_m}$$

Exercise Find input resistance and output resistance of the common-drain amplifier shown. Then, calculate the voltage gain A_v for the unloaded case and for the loaded case with $R_{sig} = 100$ kΩ and $R_L = 15$ kΩ

- $K = 0.5$ mA/V^2
- $V_t = 1.5$ V
- $\lambda = 1 / 75$

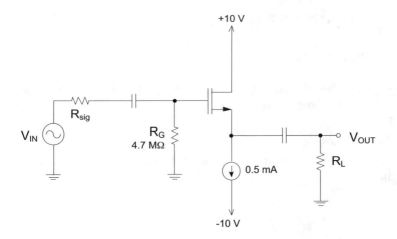

Answer The DC bias circuit is the same as the example in the common-source amplifier section so we can reuse those results, summarized here for convenience

- DCOP = (0.5 mA, 5 V)
- $V_{DSAT} = 1$ V
- $g_m = 1$ mS
- $r_o = 150$ kΩ

The input resistance $r_{in} = 4.7$ MΩ, and the output resistance $r_{out} = 1/(1\text{ m}) = 1$ kΩ.

Unloaded Gain

For the first case, $R_{sig} = 0$ and $R_L \to \infty$. Substituting,

$$A_v = + \frac{R_G}{R_{sig} + R_G} \cdot \frac{r_o \| R_L}{\dfrac{1}{g_m} + r_o \| R_L} = \frac{4.7\,\text{M}}{0 + 4.7\,\text{M}}\frac{150\,\text{k}}{1\,\text{k} + 150\,\text{k}} = 0.99\,\text{V/V}$$

Thus, the unloaded gain is close to the desired gain of +1 V/V.

Loaded Gain

$$A_v = \frac{4.7\,\text{M}}{100\,\text{k} + 4.7\,\text{M}}\frac{150\,\text{k} \| 15\,\text{k}}{1\,\text{k} + 150\,\text{k} \| 15\,\text{k}} = 0.91\,\text{V/V}$$

The loaded gain has decreased, but is still close to 1.

Before leaving the source follower, let us look at the gain equation again and write it in a slightly different way.

$$A_v = + \frac{R_G}{R_{sig} + R_G} \cdot \frac{r_o \| R_L}{\dfrac{1}{g_m} + r_o \| R_L} = \frac{r_{in}}{r_{in} + R_G} \cdot \frac{r_o \| R_L}{\dfrac{1}{g_m} + r_o \| R_L}$$

Note that again, the first term describes the voltage division between the signal generator R_{sig} and the input resistance r_{in}.

8 Applications of the Single-Transistor Amplifiers

With the gain, input resistance, and output resistance for the three topologies derived, we can begin to answer the question: which one should we choose? As in many engineering problems, the answer is "It depends on the application." We will compare the three amplifiers to the ideal voltage and current amplifiers; to simplify this comparison, we will assume $r_o \rightarrow \infty$ in all cases (Table 1).

From this, we can make a couple of general comments:

- R_G is used to provide a DC bias voltage to the gate, and since $I_G = 0$, R_G can be a large resistance.
- R_D influences the gain of the CS and CG amplifiers and cannot be too small; typical values are in the $k\Omega$ to 10's of $k\Omega$ range.
- g_m is determined by the DC operating point of the transistor. In general a higher I_D gives a larger g_m, which in turn lowers $1/g_m$, corresponding to r_{in} or r_{out} of various amplifiers.

The examples in this chapter were done with a NMOS biased in saturation at 0.5 mA, with $R_G = 4.7\ M\Omega$ and $R_D = 15\ k\Omega$. With the given transistor parameters, $g_m = 1\ mS$ in this case (Fig. 28).

The gain, r_{in}, and r_{out} are computed for the three topologies using this circuit for the unloaded gain case ($R_{sig} = 0$, $R_L \rightarrow \infty$) (Table 2).

Since a good voltage amplifier must have a large r_{in} and small r_{out}, we can make the following statements about the three topologies:

- Common Source – r_{in} is large, but r_{out} is not small.
- Common Gate – r_{in} and r_{out} do not satisfy the voltage amplifier ideals; in fact, r_{in} and r_{out} fit the requirements of a current-mode amplifier better.

Table 1 Single MOSFET amplifier equations

	A_V (V/V)	r_{in} (Ω)	r_{out} (Ω)
Common source	$-\dfrac{r_{in}}{r_{in}+R_{sig}} \cdot g_m \left(R_D \parallel R_L\right)$	R_G	R_D
Common gate	$+\dfrac{r_{in}}{r_{in}+R_{sig}} \cdot g_m \left(R_D \parallel R_L\right)$	$\dfrac{1}{g_m}$	R_D
Common drain	$+\dfrac{r_{in}}{r_{in}+R_{sig}} \cdot \dfrac{R_L}{1/g_m + R_L}$	R_G	$\dfrac{1}{g_m}$

Fig. 28 Example circuit
for amplifier comparison

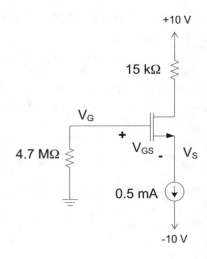

Table 2 Amplifier comparison

	A_V (V/V)	r_{in} (Ω)	r_{out} (Ω)
Common source	−15	4.7 M	15 k
Common gate	+15	1 k	15 k
Common drain	+1	4.7 M	1 k

- Common Drain – Among the three topologies, r_{in} and r_{out} are closest to that of an ideal voltage amplifier; the issue is that this circuit does not amplify (maximum gain of +1).

We can see that each amplifier has its own shortcomings, with the common-gate topology better suited to being a current-mode amplifier. In fact, in a later chapter we will look at how a common-gate amplifier can be used as a current-mode buffer (i.e., gain $A_I = +1$).

If we view these three single-transistor topologies as basic building blocks, then perhaps we can mix and match them to create a better amplifier. For example, since the source follower has the best voltage amplifier characteristics except for the gain of +1, we can use it as a voltage-mode buffer. In other words, we can use a common-source amplifier for gain, followed by a source follower to drive loads. The signal loss from the common-source output to the source follower input is minimal, as the source follower can have a very high r_{in} (Fig. 29).

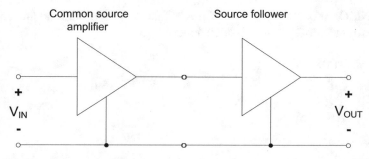

Fig. 29 An improved voltage amplifier using two stages

9   Problems

1. A voltage-mode amplifier with $A_v = -10$ is used below

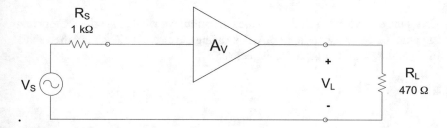

(a) For an ideal voltage-mode amplifier, what is r_{in} and r_{out} equal to?
(b) If the amplifier above has $r_{in} = 2$ kΩ and $r_{out} = 100$ Ω, what is the voltage gain V_L/V_S?

2. A current-mode amplifier with $A_I = 15$ is used below

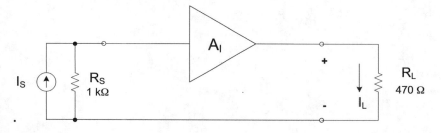

(a) For an ideal current-mode amplifier, what is r_{in} and r_{out} equal to?
(b) If the amplifier above has $r_{in} = 100$ Ω and $r_{out} = 8$ kΩ, what is the current gain I_L/I_S?

3. An NMOS transistor has $K = 2.5$ mA/V^2 and $V_t = 2$ V. Find the DC operating point of the NMOS in the circuit below.

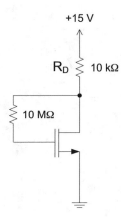

4. Using the same NMOS as the previous question, we wish to set the DC operating point of the transistor to (2 mA, 5 V) using the circuit below. Find the values of R_D and R_S that will meet this objective.

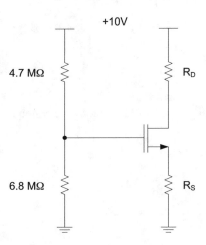

5. An NMOS transistor has a threshold voltage $V_t = 1.5$ V and $K = 30$ mA/V². Find the DC operating point of the NMOS transistor; then verify that it is in saturation.

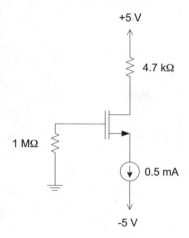

6. Using results from the previous question, and assuming the NMOS has $\lambda = 0.05$, calculate the values of g_m and r_o. Then find the gain of the common source amplifier below for the following cases:

   (a) Unloaded gain: $R_{sig} = 0$, $R_L \to \infty$
   (b) Loaded gain: $R_{sig} = 100$ kΩ, $R_L = 15$ kΩ

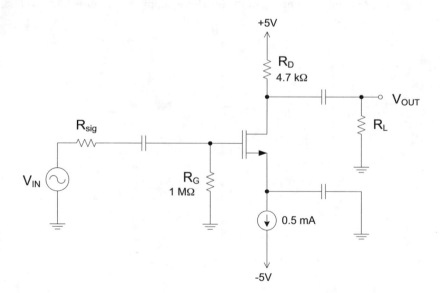

7. An NMOS transistor has a threshold voltage $V_t = 2$ V and $K = 10$ mA/V^2. Find the voltage gain A_v, r_{in}, and r_{out} of the following amplifier. Assume that $\lambda = 0$.

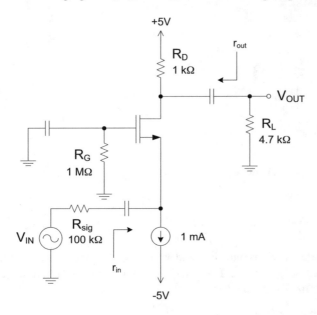

8. An NMOS transistor has a threshold voltage $V_t = 2$ V and $K = 10$ mA/V^2. Find the voltage gain A_v, r_{in}, and r_{out} of the following amplifier. Assume that $\lambda = 0.05$.

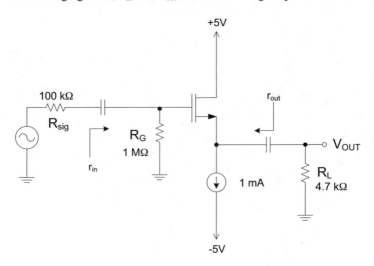

9. Find the DC operating point of the NMOS below and the gain of the common source amplifier. The parameters of the transistor are $K = 20$ mA/V^2, $V_t = 2$ V and $\lambda = 0.1$. You may ignore the effect of the 10 MΩ resistor in the AC analysis.

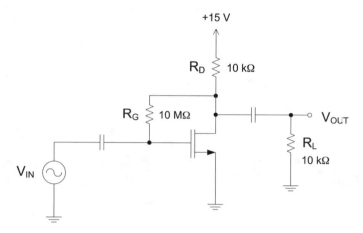

10. Design a MOSFET-based amplifier with $r_{in} \geq 1$ MΩ, $r_{out} \leq 50$ Ω, and a voltage gain of 20 V/V. The gain can be either inverting or non-inverting. NMOS transistors are available with $K = 20$ mA/V^2, $V_t = 2$ V, and $\lambda = 0.1$. Design the circuit using a single polarity 12 V supply.

11. Tinkercad®: V_t and K of the small signal NMOS transistor were found in a previous chapter's exercises. Placing this NMOS in the bias circuit shown below, simulate the DC operating point.

Small Signal
nMOS Transistor

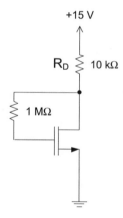

12. Tinkercad®: Configure the DC bias circuit from the previous question into a common source amplifier. Set the signal generator to output a 1 kHz, 10 mV$_{pp}$ sine wave. Measure the input and output using the scope, and estimate the voltage gain.

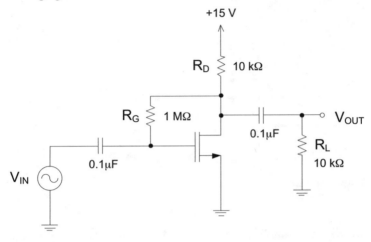

PMOS and CMOS

Just like the NPN transistor which has a counterpart in the PNP, the NMOS transistor has a sibling in the PMOS transistor. NMOS and PMOS transistors can be manufactured in the same integrated circuit, resulting in the modern CMOS (complementary metal oxide semiconductor) technology that is widely used.

1   PMOS

Like the NMOS, the PMOS has a three-terminal and a four-terminal symbol. Common to both symbols are the terminals *gate*, *drain*, and *source*. The four-terminal symbol shows an additional, intrinsic connection called the *bulk* or *body*. In most discrete transistors, the *bulk* and *source* terminals are shorted together.

Note that the PMOS symbol differs from the NMOS symbol in one aspect: the arrow on the source terminal is reversed. Also, the author has chosen to draw the PMOS symbol with the source terminal on top, for the same reason as the PNP: to turn on the PMOS, the source terminal must have a higher voltage than the gate terminal (Fig. 1).

Without going into the physics of it, note that we typically want to connect the PMOS source terminal to the highest voltage of a system (e.g., V_{DD}). Conversely, we typically connect the NMOS source terminal to the lowest voltage in a system (0 V or $-V_{SS}$) (Fig. 2).

- In many books and datasheets, the PMOS threshold voltage is specified as negative. This all makes sense using their convention and notation. However, in this book we will use the magnitude of the threshold voltage and denote it as $|V_{tp}|$.
- NMOS and PMOS devices, even in the same integrated circuit, typically have different threshold voltages (V_{tn} and $|V_{tp}|$).

© Springer Nature Switzerland AG 2022
C. Siu, *Electronic Devices, Circuits, and Applications*,
https://doi.org/10.1007/978-3-030-80538-8_7

Fig. 1 PMOS symbol (**a**)
3-terminals (**b**) 4-terminals

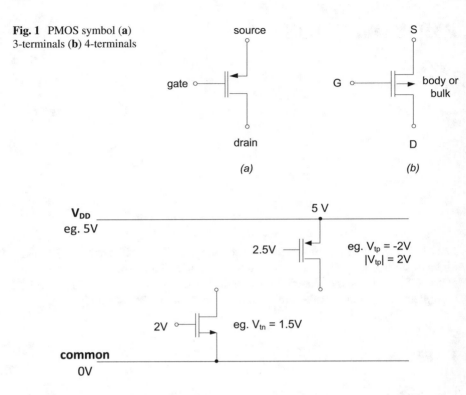

(a)   *(b)*

Fig. 2 Typical usage of NMOS and PMOS transistors

- For the NMOS to turn on, the gate must be higher than the source by at least the threshold voltage ($\mathbf{V_{GS}} > \mathbf{V_{tn}}$); in the above diagram, $V_{GS} = 2$ V and is greater than V_{tn} by 0.5 V, yielding $V_{DSAT} = 0.5$ V.
- For the PMOS to turn on, the source must be higher than the gate by at least the threshold voltage ($\mathbf{V_{SG}} > |\mathbf{V_{tp}}|$); in the above diagram, $V_{SG} = 2.5$ V and is greater than $|V_{tp}|$ by 0.5 V, yielding $V_{DSAT} = 0.5$ V.
- NMOS and PMOS transistors also have different values of K, which will be denoted as $\mathbf{K_n}$ and $\mathbf{K_p}$, respectively. The NMOS and PMOS equations are shown below for comparison; they are similar but do pay attention to the subscripts (for NMOS, we use V_{GS}; for PMOS, we use V_{SG}).

$I_D = K_n(V_{GS} - V_{tn})^2$	NMOS in saturation
$I_D = 2K_n \left[\left(V_{GS} - V_{tn} \right) \times V_{DS} - \dfrac{V_{DS}^2}{2} \right]$	NMOS in triode
$V_{DSAT} = V_{GS} - V_{tn}$	NMOS
$I_D = K_p(V_{SG} - |V_{tp}|)^2$	PMOS in saturation

$I_D = 2K_p \left[\left(V_{SG} - \left\| V_{tp} \right\| \right) \times V_{SD} - \dfrac{V_{SD}^2}{2} \right]$	PMOS in triode
$V_{DSAT} = V_{SG} - \left\| V_{tp} \right\|$	PMOS

Exercise Determine whether the PMOS is on or off when V_{in} changes between 0 to 5 V. If the PMOS is on, determine its operating region and the DC operating point:

- $K_p = 1 \text{ mA}/V^2$
- $\left| V_{tp} \right| = 1.5 \text{ V}$

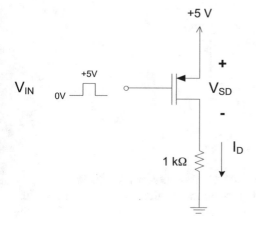

Answer If the input is at 5 V, then $V_{SG} = 0$ V, and the PMOS is off.

If the input is at 0 V, then $V_{SG} = 5$ V, and since this exceeds the threshold voltage, the PMOS is on. We will assume the device is in saturation, and check to see if this assumption is valid.

$$I_D = K_p \left(V_{SG} - \left| V_{tp} \right| \right)^2 = (1m)(5-1.5)^2 = 12.25 \text{ mA}$$

$$V_D = (12.25m)(1k) = 12.25 \text{ V}$$

The calculated drain voltage is 12.25 V, which is higher than the supply voltage of +5 V. It is not possible to have a node voltage higher than V_{DD} in this circuit, so our assumption is wrong: the PMOS is in the triode region. We will use the approximate method, which drops the V_{SD}^2 term in the equation

$$I_D = 2K_p\left[\left(V_{SG} - |V_{tp}|\right) \times V_{SD} - \frac{V_{SD}^2}{2}\right] \approx 2K_p\left(V_{SG} - |V_{tp}|\right) \times V_{SD}$$

We can now model the PMOS in triode with an equivalent resistance between source and drain:

$$R_{SD(ON)} = \frac{V_{SD}}{I_D} = \frac{1}{2K_p\left(V_{SG} - |V_{tp}|\right)} = \frac{1}{2(1m)(5-1.5)} = 143\,\Omega$$

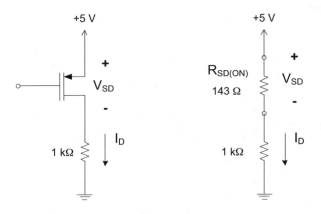

Using Ohm's law and voltage division, the DCOP is 4.37 mA, 0.626 V.

The exact method can also be used. Though the details are not shown here, the technique is the same as that used in an earlier NMOS exercise. The DCOP using the exact method is 4.32 mA, 0.685 V.

2   CMOS

In CMOS technology, both PMOS and NMOS transistors are used to create specific circuit functions. Often, CMOS is used to create logic gates and complex digital functions, resulting in the microprocessors and microcontrollers that we have today. We will study some simple logic gates constructed from CMOS, starting with the logic inverter.

The CMOS Inverter

A logic inverter can be constructed by using one NMOS and one PMOS transistor, connected as shown below. To understand how this circuit works, suppose that $V_{DD} = 5$ V, $V_{tn} = 1.5$ V, and $|V_{tp}| = 2$ V (Fig. 3).

If the input X is at logic 0 (0 V), then the PMOS is on and the NMOS is off. The PMOS M2 will enter the triode region, acting like a pull-up resistor to V_{DD}. The output is therefore equal to logic 1 (5 V).

If the input X is at logic 1 (5 V), then the PMOS is off and the NMOS is on. The NMOS M1 will be in triode and act like a pull-down resistor to ground. The output is therefore logic 0 (0 V).

Summarizing this information in a table, we can clearly see that this circuit is a logic inverter (Table 1).

If we assume that the input voltage has been applied for long enough such that the output has settled, the inverter can be represented in one of two states (Fig. 4).

Note that whether the input is logic 0 or logic 1, one transistor is off, and there is no path for current to flow from V_{DD} to ground. The DC current consumption is 0, and this is the big advantage of CMOS over the BJT-based logic family TTL (Transistor Transistor Logic): *the static current consumption of CMOS is 0.*

Alas, the "No Free Lunch" principle still applies to CMOS. When the inverter is switching, there is a brief period of time when both transistors are on, causing power to be dissipated; this is called the short circuit current. As shown in Fig. 5, when the

Fig. 3 The CMOS inverter

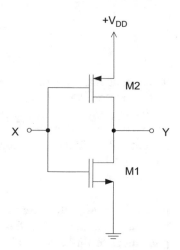

Table 1 CMOS inverter operation

X	NMOS M1	PMOS M2	Y
0 V	Off	On	+5 V
+5 V	On	Off	0 V

Fig. 4 Static equivalent
circuits for a CMOS
inverter

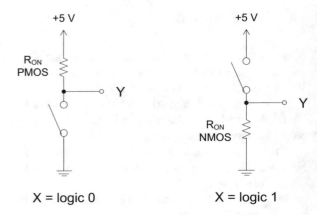

Fig. 5 CMOS inverter
during an input transition

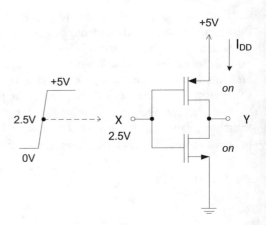

input changes from 0 to 5 V, there is a brief moment when the input is around 2.5 V
and both transistors are on.

Each time the inverter switches, a current spike is drawn from the power supply.
The wires connecting the circuit to the power supply have self-inductances, and the
current spike will induce voltage changes across these inductances, creating *supply
noise* (Fig. 6).

To reduce the switching noise on V_{DDi} and V_{SSi}, it is a common practice to use
bypass capacitors in circuit board design. As shown in Fig. 7, a capacitor is placed
physically close to the inverter's power and ground pins. When the inverter switches
and draws a current spike, the bypass capacitor acts like a local reservoir that supplies most of this charge. Since less of the current spike is coming through the
power supply inductance, there is less supply and ground noise generated.

Fig. 6 Current drawn by a
CMOS inverter during a
transition

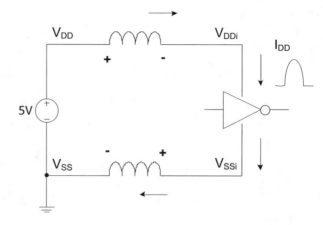

Fig. 7 Bypass capacitors
for reducing power and
ground noise

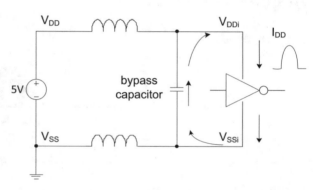

Propagation Delay

To understand the causes of propagation delay, we can look at the two cascaded
inverters below. Suppose that initially, the input X is V_{DD}, and hence the node Y is at
0 V (Fig. 8).

Since inverter #1 drives inverter #2, what kind of load does the input of inverter
#2 present? We know that the DC gate current is 0, so it is an open circuit at DC. Is
it also an open circuit for AC or time-varying signals?

The answer can be deduced by looking at the physical structure of a MOSFET,
for example, an NMOS as shown below (Fig. 9).

In the figure, the source and bulk are shorted together. As we apply a positive
voltage to the gate, we can visualize a number of positive charges being on the gate
metal. Under the gate metal is an insulator (silicon dioxide), so those positive
charges cannot move from the gate into the substrate. Finally, if $V_{GS} > V_{tn}$, then there
are sufficient positive charges to attract negative charges to the surface of the semiconductor, beneath the silicon dioxide. These negative charges "connect" the drain
to source, making current flow from drain to source possible.

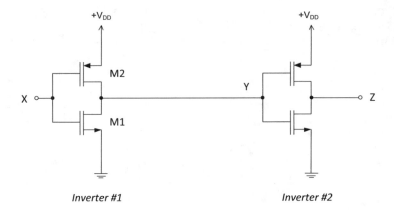

Inverter #1   Inverter #2

Fig. 8 CMOS inverters in cascade

Fig. 9 NMOS gate
capacitance

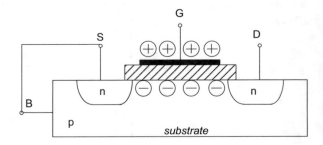

What we have just described here, the conductor-insulator-conductor sandwich storing charge … this is the characteristics of a capacitor! The gate of a MOSFET is capacitive, and it represents a parasitic capacitance that is intrinsic to the MOSFET. Extending this one step further, we can conclude that the input of a logic inverter is also capacitive. We can now redraw the cascaded inverter schematic and replace the second inverter with a capacitor (Fig. 10).

Initially at time t1 the input X is 5 V and the output Y is 0 V. At time t2, X is in the midst of changing. Imagine that X has just reached 0 V, so the PMOS is on and the NMOS is off. Output Y does not jump to +5 V right away, as the capacitor C_{in}'s voltage cannot change instantaneously. Specifically, we can visualize the PMOS acting like a pull-up resistor to V_{DD}, charging C_{in} to +5 V after some time (Fig. 11).

Visually, due to the slow rise time at the output, it appears that there is a delay from the input X to the output Y, denoted as the *propagation delay* t_{PD}. Generally, this is one key factor behind propagation delay: the charging and discharging of CMOS inputs. Note that this charging and discharging consumes power and increases the power dissipation of CMOS gates beyond that of the short circuit current.

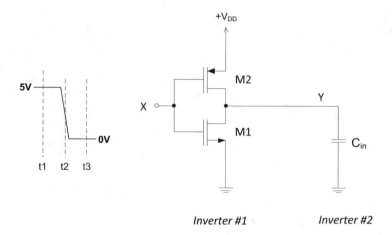

Fig. 10 Inverter #1 driving input capacitance of Inverter #2

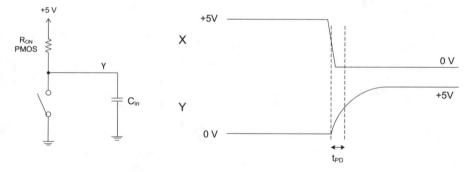

Fig. 11 Propagation delay due to charge & discharge of input capacitance

CMOS NAND Gate

We can now examine other logic gates built with CMOS technology. Shown below is a circuit for a NAND gate, built with two NMOS and two PMOS transistors (Fig. 12).

Recall that the Boolean equation for a NAND gate is

$$Y = \overline{(X1 \cdot X2)}$$

In this circuit, any transistor turned on is driven into the triode region. For the output Y to be pulled to ground, note that both NMOS M3 and M4 must be turned on. M3 and M4 are both on if $X1 = X2 = V_{DD} = $ logic 1, during which the PMOS M1 and M2 are off. Hence for $X1 = 1$ and $X2 = 1$, $Y = 0$ as expected for a NAND gate.

Fig. 12 CMOS
NAND gate

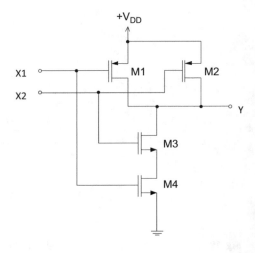

Fig. 13 CMOS NAND
gate with X1 = 1, X2 = 0

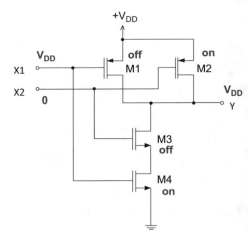

For the inputs X1 = 1 and X2 = 0, we can quickly deduce which transistors are on or off (Fig. 13).

The reader can try other input combinations and verify that this circuit performs the NAND function.

CMOS Transmission Gate

A transmission gate can be thought of as an analog switch. It can be used to pass or stop an analog signal between two points. The symbol and circuit for a transmission gate is shown below (Fig. 14).

A transmission gate consists of one NMOS and one PMOS transistor. When Enable = 0, both transistors are off and the output is high impedance. When

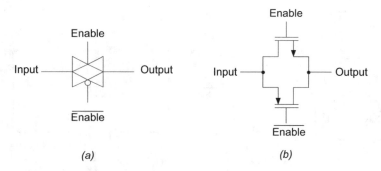

Fig. 14 CMOS transmission gate (a) symbol (b) circuit

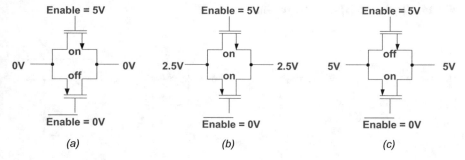

Fig. 15 CMOS transmission gate in enabled state, for various input voltages

Enable = 1, then both transistors may be on depending on the input voltage. To illustrate this point, suppose that logic $1 = 5$ V, $V_{tn} = 1.5$ V, and $|V_{tp}| = 2$ V. If the input varies between 0 V and 5 V, the transistors go through different states as shown below (Fig. 15).

To simplify the analysis, we will assume that when Enable = 1, then the input and output voltages are the same. First, when the input = 0 V, note that $V_{GSN} = 5$ V and $V_{SGP} = 0$ V. Hence, the NMOS is on and in triode, and the PMOS is off as shown in (a).

In (b) when the input = 2.5 V, $V_{GSN} = 2.5$ V and $V_{SGP} = 2.5$ V, and both transistors are in triode. Finally, in (c) when the input = 5 V, $V_{GSN} = 0$ V and $V_{SGP} = 5$ V, and so the NMOS is off and the PMOS is on. When enabled, the CMOS transmission gate has at least one transistor on so that continuity is maintained between the input and the output.

CMOS Latch

A basic element in many digital systems is memory, with flip-flop circuits forming one kind of memory. Flip-flops can be constructed with latches, of which one type is shown below. Functionally, when CLK = 1, then Q = D. If CLK = 0, then the last value of D is stored, being either a 0 or 1 (Fig. 16).

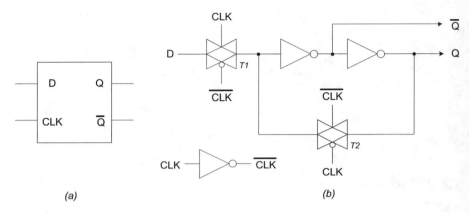

Fig. 16 CMOS latch (a) symbol (b) circuit

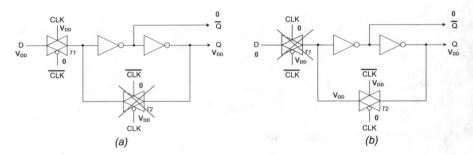

Fig. 17 CMOS latch (a) sample the input D (b) hold mode

The operation of this circuit can be explained as follows. When CLK = 1, the transmission gate T1 is closed and T2 is open. Hence, D is the input into the two cascaded inverters, resulting in Q = D.

When CLK = 0, T1 is now opened and T2 is closed. D is disconnected from the input, and a feedback path is formed around the two inverters. Note that with the feedback, the two inverters can hold its previous states, storing the last value of D (Fig. 17).

3   Problems

1. A PMOS transistor has $K_p = 10$ mA/V^2 and $|V_{tp}| = 2$ V. Find the DC operating point for the following cases:

   (a) $V_B = 2.5$ V
   (b) $V_B = 0$ V

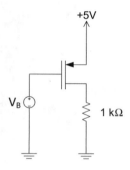

2. What is the reason for using bypass capacitors?
3. The circuit below is a CMOS logic gate. Specify which logic function it implements by compiling a logic table.

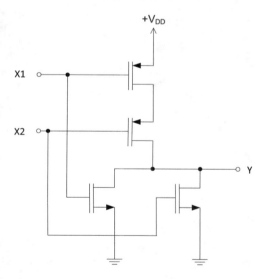

4. The circuit below is a CMOS logic gate. Specify which logic function it imple-
    ments by compiling a logic table.

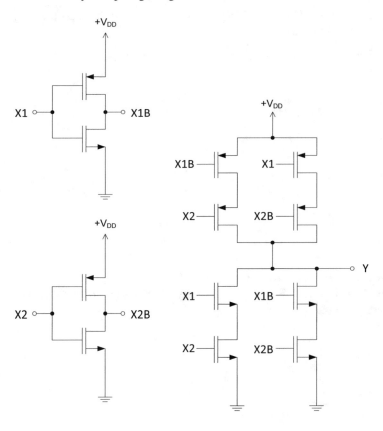

5. The following circuit implements a logic gate known as a tristate buffer. Determine the logic table for this gate.

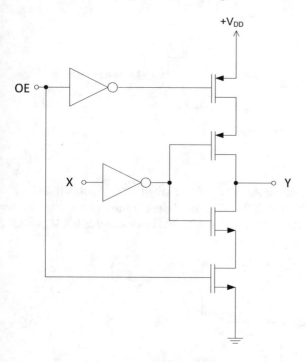

6. A CMOS inverter uses a NMOS transistor with $K_n = 20$ mA/V^2 and $V_{tn} = 1.5$ V and a PMOS transistor with $K_p = 10$ mA/V^2 and $|V_{tp}| = 2$ V. The inverter is driving a 2 pF load, operating with a 5 V power supply. Estimate the propagation delay of the inverter under this load condition.

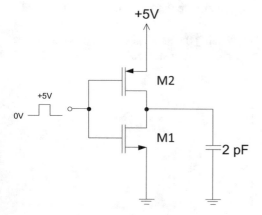

7. A CMOS transmission gate uses an NMOS transistor with $K_n = 20$ mA/V^2 and $V_{tn} = 1.5$ V and a PMOS transistor with $K_p = 10$ mA/V^2 and $|V_{tp}| = 2$ V. Estimate the resistance of the transmission gate when enabled, for the following input

voltages: 0 V, 2.5 V, and 5 V. *Hint: note that with no load at the output, the input and output voltages are the same when the transmission gate is enabled.*

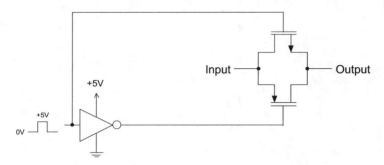

8. Tinkercad®: Measure the threshold voltage of the small signal PMOS transistor using the circuit below. We will define V_t as the highest source-gate voltage that has a drain current of zero; test your circuit with V_{SG} increments of 0.1 V to find the threshold voltage.

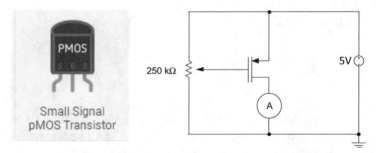

9. Tinkercad®: Measure I_D for multiple values of V_{SG}. Using this and the threshold voltage found in the previous question, estimate the value of K for this PMOS transistor.

10. Tinkercad®: Simulate the circuit below using the small signal PMOS transistor, and record I_D and V_{SD}. Compare this result against prediction using K and V_t obtained in the previous questions.

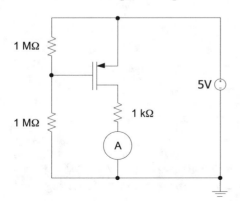

11. Tinkercad®: Construct a CMOS inverter using the small signal NMOS and PMOS transistors, and verify its operation.
12. Tinkercad®: Construct a CMOS transmission gate using the small signal NMOS and PMOS transistors. For the inverter in the Enable signal path, use a 74HC04 Hex Inverter IC; search the Internet for the datasheet and pinout. The power supply for the 74HC04 is 4 × AA batteries, which provides 6 V.

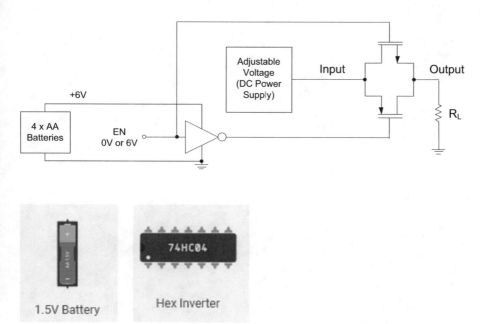

First, test the transmission gate by applying 3 V to the input and measuring the output voltage with $R_L = 1$ MΩ. Record the output voltage when the gate is enabled and disabled.

Next, change the load to $R_L = 1$ kΩ and enable the gate. Measure the voltage drop and the current through the transmission gate. Estimate its resistance with these measurements for input voltages of 1 V, 3 V, and 5 V.

Frequency Response

The analysis in a previous chapter has yielded the gain equations for three basic amplifier topologies. To convert this voltage gain to dB, we can use the following (Fig. 1):

$$A_{V(dB)} = 20 \cdot \log_{10} \left| A_V \right| = 20 \cdot \log_{10} \left| \frac{v_{out,PK}}{v_{in,PK}} \right|$$

If we change the input frequency over a limited range, we may expect the gain to stay constant. Should we expect this gain to stay constant if we keep increasing the frequency? In a real amplifier, the gain will drop beyond some frequency limit, often denoting the bandwidth of the amplifier (Fig. 2).

We will look at the most common metric for bandwidth: the 3 dB bandwidth.

1 The 3 dB Bandwidth

The frequency range that an amplifier can provide constant gain is called the passband. In a typical design, the passband should cover all signals of interest in the application. For example, the human voice contains frequencies up to 4 kHz, and thus an amplifier for this application must have a passband extending up to 4 kHz (Fig. 3).

The 3 dB bandwidth, 3 dB cutoff frequency, or half power point are different names for the same metric: the frequency f_{3dB} at which the amplifier gain is 3 dB lower than the passband gain. We can also show that at this frequency, the output power is halved relative to the passband.

One of the simplest circuit topologies, the RC circuit, is also known as a low-pass filter. The lower frequencies within the passband experience a constant gain of 1 V/V or 0 dB through the filter. The higher frequencies above f_{3dB} are attenuated by

© Springer Nature Switzerland AG 2022
C. Siu, *Electronic Devices, Circuits, and Applications*,
https://doi.org/10.1007/978-3-030-80538-8_8

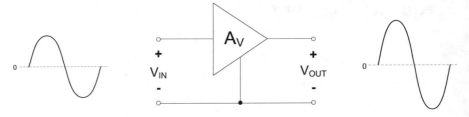

Fig. 1 Voltage amplifier input & output

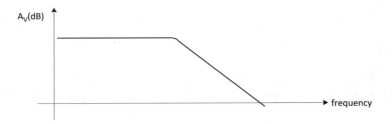

Fig. 2 Gain of a typical amplifier versus frequency

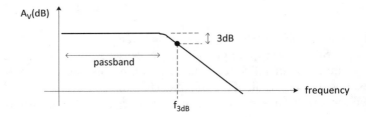

Fig. 3 3 dB bandwidth definition

Fig. 4 First order RC circuit

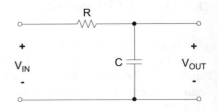

the filter. We will find an expression for the 3 dB bandwidth of an RC circuit, and it is an expression that is worth memorizing given its usage in electrical engineering (Fig. 4).

We can find the voltage gain by applying the voltage divider rule, where the AC voltage V_{IN} is divided between resistance R and the capacitive reactance $1/(j\omega C)$.

$$A_\mathrm{v} = \frac{v_\mathrm{out}}{v_\mathrm{in}} = + \frac{1/j\omega C}{R + 1/j\omega C} = \frac{1}{1 + j\omega RC}$$

To express the gain in dB, we must find the magnitude of this expression.

$$|A_\mathrm{v}| = \left| \frac{1/j\omega C}{R + 1/j\omega C} \right| = \frac{1}{\sqrt{1 + (\omega RC)^2}}$$

$$A_\mathrm{v(dB)} = 20 \cdot \log_{10} \frac{1}{\sqrt{1 + (\omega RC)^2}}$$

If we equate the last expression to −3 dB and solve for ω, we will find the equation for the 3 dB cutoff frequency of a RC circuit:

$$\omega_\mathrm{3dB} = \frac{1}{RC}$$

$$f_\mathrm{3dB} = \frac{1}{2\pi RC}$$

2 Bode Plot

A Bode plot is a plot of the frequency response where the horizontal axis is the log of frequency, and the vertical axis is the magnitude in decibels. This is a common format for graphing frequency responses, and if you have heard of transfer functions, then a Bode plot can be sketched quickly using transfer functions (Fig. 5).

In this introduction, we will use the Bode plot to display the characteristics of the RC circuit we have just studied.

Example Calculate the 3 dB bandwidth of the circuit below. Next, sketch the Bode plot for this circuit.

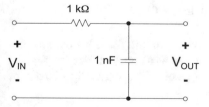

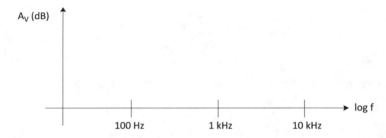

Fig. 5 Bode plot definition

Answer Using equation $f_{3dB} = (2\pi RC)^{-1}$,

$$f_{3dB} = \frac{1}{2\pi(1k\Omega)(1nF)} = 159\,kHz$$

Recall the gain equation derived earlier. Since we now know that $\omega_{3dB} = 1/RC$, we can write this equation as shown

$$|A_v| = \frac{1}{\sqrt{1+(\omega RC)^2}} = \frac{1}{\sqrt{1+\left(\dfrac{\omega}{\omega_{3dB}}\right)^2}} = \frac{1}{\sqrt{1+\left(\dfrac{f}{f_{3dB}}\right)^2}}$$

At input frequencies much lower than f_{3dB}, the term $(f/f_{3dB})^2$ is very small compared to 1. Hence, the denominator is approximately 1, and the passband gain $A_V \sim 1$ V/V or 0 dB.

For input frequencies much higher than f_{3dB}, then the term $(f/f_{3dB})^2$ is large and the denominator is approximately f/f_{3dB}, and $A_V = 1/(f/f_{3dB}) = f_{3dB}/f$. As f becomes larger, A_V gets smaller and goes to 0 as $f \to \infty$. High-frequency signals are attenuated and do not appear at the output. Since we are creating a Bode plot, the vertical axis is in dB, and if $f \gg f_{3dB}$,

$$A_{v(dB)} = 20 \cdot \log_{10}\frac{1}{\sqrt{1+\left(\dfrac{f}{f_{3dB}}\right)^2}} \approx 20 \cdot \log_{10}\left(\frac{f_{3dB}}{f}\right)$$

$$= 20 \cdot \log_{10}(f_{3dB}) - 20 \cdot \log_{10}(f)$$

In a Bode plot, the horizontal axis is $\log_{10}(f)$. Given a circuit, f_{3dB} is set by the component values, and hence $20 \cdot \log_{10}(f_{3dB})$ can be viewed as a constant. For $f \gg f_{3dB}$, the expression above indicates the Bode plot has a slope of -20, specifically -20 dB for one decade increase in frequency.

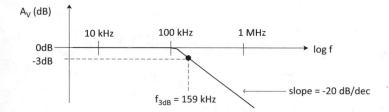

If the input frequency is at f_{3dB}, then the output is 3 dB lower than what it would be in the passband. For the RC circuit, the passband gain is 0 dB, and hence the gain at f_{3dB} is −3 dB. In term of voltage, the scaling factor at f_{3dB} is

$$A_{v(dB)} = 20 \cdot \log_{10}\left(\frac{V_{out}}{V_{in}}\right) = -3\,dB$$

$$\frac{V_{out}}{V_{in}} = 10^{-3/20} = 0.707$$

Exercise A 5 V peak sine wave at a frequency equal to f_{3dB} is applied to the input of this circuit. What is the output equal to?

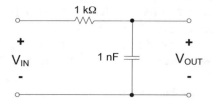

Answer In the passband, this circuit has a gain of +1 V/V, and hence the output is 5 V peak. If the frequency is increased to f_{3dB}, then the output is $5\,V \cdot 0.707 = 3.54\,V$ peak.

3   Amplifier Bandwidth

The fact that a RC circuit has a limited bandwidth provides us with insight on why electronic amplifiers have finite bandwidths. Beginning with an ideal voltage amplifier ($r_{in} \to \infty$, $r_{out} = 0$), we add a capacitor C_{in} at the input; C_{in} may include the intrinsic capacitances of MOSFETs, for example.

Due to the signal generator R_{sig}, a RC circuit is formed at the input, with 3 dB bandwidth given by (Fig. 6)

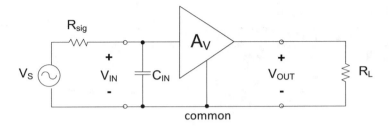

Fig. 6 Effect of input capacitance on bandwidth

$$f_{3dB} = \frac{1}{2\pi R_{sig} C_{in}}$$

The Miller Effect

Instead of C_{in}, what if there is a capacitor C_M connected between the amplifier input and output? We will see later how this may happen, but for now we wish to answer the question: what is the 3 dB bandwidth of this circuit? (Fig. 7)

While we may be tempted to answer $(2\pi \cdot R_{sig} C_M)^{-1}$, there is a key difference between the previous and present scenarios. Note that one terminal of C_{in} is connected to common, whereas neither terminals of C_M are connected to common. An interesting effect occurs if the amplifier is inverting, meaning that A_V has a negative value. To make this explicit, we will set $A_V = -A$, where A is a positive real number greater than one (Fig. 8).

To solve this problem, we will take a more general approach by finding the input impedance Z_{in}. As shown in Fig. 8a, we have a black box that has some impedance across the two terminals. Applying a test voltage Δv, we find the resulting change in current Δi through those terminals. The ratio $\Delta v/\Delta i$ will yield the impedance Z_{in}.

Suppose that inside the black box there is an ideal voltage amplifier with inverting gain and a capacitor C connected between the input and output. As shown in Fig. 8b, a test voltage Δv is applied, and the amplifier output is $-A \cdot \Delta v$. Note that because the capacitor is connected between the input and output, the voltage it sees is $\Delta v - (-A \cdot \Delta v) = \Delta v \cdot (1 + A)$. The voltage change across the capacitor is hence bigger than the applied input, since $A > 1$.

In addition, by applying the test voltage Δv, we need to find Δi for calculating Z_{in}. Since we are using an ideal voltage amplifier, the input resistance is infinite. All current Δi must then flow through the capacitor. Given that the capacitive reactance is $1/(j\omega C)$, we can use Ohm's law to find Δi:

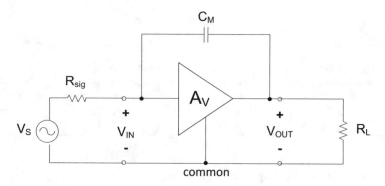

Fig. 7 Effect of capacitance between the input and output

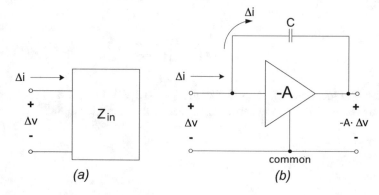

Fig. 8 Miller effect (**a**) impedance measurement concept (**b**) analysis

$$\Delta i = \frac{\Delta v \cdot (1 + A)}{1 / j\omega C} = \Delta v \cdot (1 + A) \cdot j\omega C$$

$$Z_{in} = \frac{\Delta v}{\Delta i} = \frac{1}{(1 + A) \cdot j\omega C} = \frac{1}{j\omega C (1 + A)} = \frac{1}{j\omega C_{eff}}$$

The input impedance Z_{in} is still capacitive, but the capacitance is $C_{eff} = C(1 + A)$. In other words, the capacitor C connected between the input and output looks a lot bigger due to the inverting amplifier. There is a large voltage change across C even for a small Δv, and the large voltage change creates a large current change. This is the physical mechanism by which the capacitance seems bigger and is called the *Miller effect* (Fig. 9).

Figure 9a shows the setup that we have been analyzing, with the capacitor C_M connecting across the amplifier input and output. Figure (b) shows how we can represent C_M in an equivalent manner, by having $C_{eff} = C_M \cdot (1 + A)$ connected

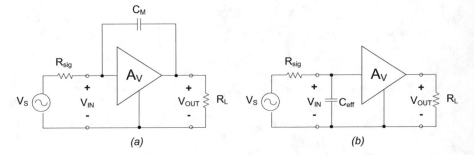

Fig. 9 Miller effect (**a**) Miller capacitor (**b**) Equivalent input capacitance

between the input and common. We can now find the 3 dB bandwidth by inspection from Fig. 9b

$$f_{3dB} = \frac{1}{2\pi R_{sig} C_M \left(1 + A\right)}$$

Exercise An ideal voltage amplifier with $A_V = -10$ V/V has a 1 nF capacitor connected as shown. What is the capacitance seen at the amplifier input?

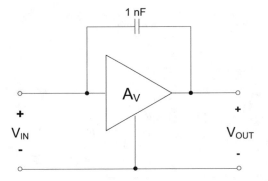

Answer Since the capacitor is connected between the input and output of an inverting amplifier, the Miller effect applies, and $C_{eff} = 1$ nF·$(1 + 10) = 11$ nF.

Exercise An ideal voltage amplifier with $A_V = -10$ V/V has a 1 nF capacitor connected as shown. What is the capacitance seen at the amplifier input?

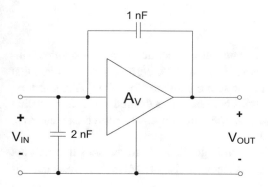

Answer As in the prior example, the Miller effect causes the 1 nF to appear as an input capacitance of 11 nF. This is in parallel to the additional 2 nF, and hence $C_{\text{eff}} = 13$ nF.

Exercise An ideal voltage amplifier with $A_V = -10$ V/V is driven by a signal with a source resistance of 1 kΩ. Calculate the 3 dB bandwidth of the response from V_{in} to V_{out}.

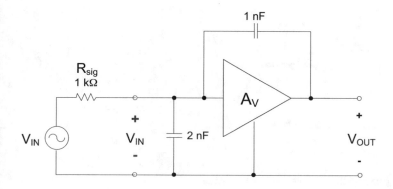

Answer In the previous example, C_{eff} was found to be 13 nF. This forms a low pass filter with R_{sig}, resulting in a $f_{\text{3dB}} = (2\pi \cdot 1 \text{ k}\Omega \cdot 13 \text{ nF})^{-1} = 12.2$ kHz.

4   MOSFET High-Frequency Model

The NMOS small signal model presented in an earlier chapter assumes that the operating frequency is low. To quantify if a frequency can be considered high or low, we must now review the structure of an NMOS transistor.

MOSFET Intrinsic Capacitances

As shown previously, between the gate and the semiconductor is an insulator (silicon dioxide), which means that the gate current at DC is zero. It is also evident that the gate contact forms a capacitor with the semiconductor. To switch on the NMOS, we must charge this gate capacitance until V_{GS} reaches its desired value. In modeling this capacitance, we can split it into two components: a gate-source capacitor C_{gs} and a gate-drain capacitor C_{gd} (Fig. 10).

For an NMOS in saturation, C_{gs} is much larger than C_{gd} for reasons that will not be explained here. If we review the datasheet of a discrete MOSFET, these capacitances typically go by different names:

- Input capacitance = $C_{iss} = C_{gs} + C_{gd}$
- Reverse transfer capacitance = $C_{rss} = C_{gd}$

For example, a commercially available MOSFET rated for 200 A with $R_{DS(ON)} = 1.5$ mΩ has a $C_{iss} = 8800$ pF and $C_{rss} = 40$ pF. From this, we can deduce that $C_{gs} = 8760$ pF and $C_{gd} = 40$ pF.

We can now augment the NMOS small signal model to include these intrinsic capacitances, resulting in the high-frequency model (Fig. 11):

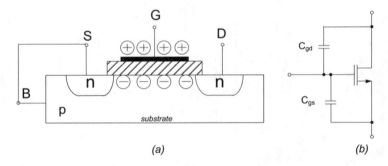

(a) (b)

Fig. 10 NMOS gate capacitance (**a**) physics (**b**) model

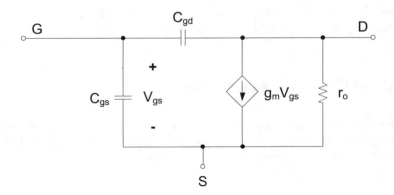

Fig. 11 NMOS high frequency small signal model

5 Bandwidth of Common Source Amplifier

With the necessary theory in place, we can now analyze the bandwidth of a common source amplifier (Fig. 12).

- The flat portion of the frequency response is the passband. The gain within the passband is referred to as the midband gain, and this gain can be predicted using the analysis we performed in an earlier chapter
- Our analysis assumed that the capacitors C1, C2, and C3 can be treated as AC short circuits. This assumption is valid if the capacitors are large enough at the frequencies of interest. As we decrease the input frequency, at some point this assumption is not valid, and our gain equation no longer holds. Intuitively, we can see that close to DC, the capacitors will be open circuits, and hence the gain will tend toward zero.
- Implicitly assumed in our analysis is that the MOS transistor is infinitely fast. A real MOSFET has a finite speed, with the parasitic capacitances C_{gs} and C_{gd} discussed earlier. This sets a limit on the upper cutoff frequency f_H.

In the following discussion, we will create an estimate for f_H. We start by redrawing the schematic by replacing the capacitors with an AC short circuit; in other words, we are drawing the circuit as the AC input signal would see it (Fig. 13).

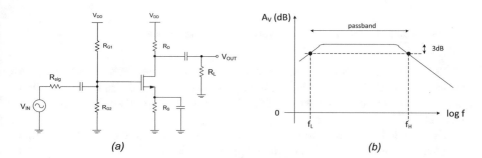

Fig. 12 Common source amplifier (**a**) circuit (**b**) frequency response

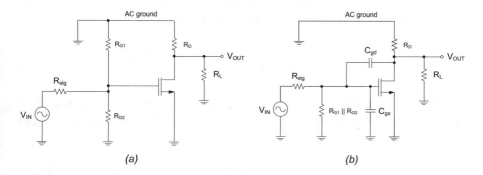

Fig. 13 CS amplifier (**a**) AC equivalent schematic (**b**) with MOS capacitors

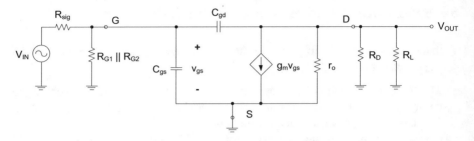

Fig. 14 AC equivalent circuit of CS amplifier with MOS capacitors

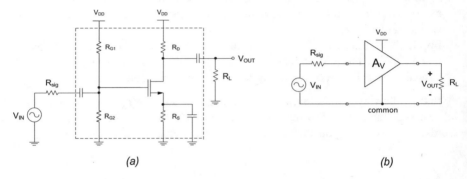

Fig. 15 CS amplifier (**a**) circuit (**b**) ideal voltage amplifier representation

Next, we add the NMOS intrinsic capacitors to the schematic. Note that C_{gd} is subject to the Miller effect, since it is connected between the amplifier input and output. We can also replace the NMOS with its high-frequency small signal model (Fig. 14).

The detailed analysis of this equivalent circuit is shown in the appendix of this chapter. We will take a more simplified approach that will also give us insight into which elements are limiting the bandwidth.

In the simplified approach, we will assume that the common source amplifier is an ideal voltage amplifier (it is acknowledged that the common source amplifier is far from ideal for output resistance), with a voltage gain as derived before (Fig. 15):

$$A_v = \frac{v_{out}}{v_{in}} = -\frac{R_{G1} \| R_{G2}}{R_{sig} + R_{G1} \| R_{G2}} \cdot g_m \left(r_o \| R_D \| R_L \right)$$

By attaching C_{gs} and C_{gd} to the amplifier model, we can now see the Miller effect applies to C_{gd}. Using the equation derived earlier, we can find an equivalent input capacitance due to C_{gd} (Fig. 16).

$$C_{in} = C_{gs} + \left(1 - A_v \right) \cdot C_{gd}$$

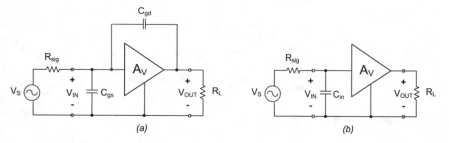

Fig. 16 CS amplifier model (**a**) with MOS caps (**b**) equivalent caps

At the input, R_{sig} and C_{in} form a low pass filter which limits the amplifier bandwidth. The resistors R_{G1} and R_{G2} can be ignored provided that $R_{\text{sig}} \ll (R_{\text{G1}} \parallel R_{\text{G2}})$.

Example For the common source amplifier below, estimate the upper cutoff frequency f_H for the following parameters:

- $K = 0.5 \text{ mA/V}^2$
- $V_t = 1.5 \text{ V}$
- $\lambda = 1/75$
- $C_{\text{gs}} = 10 \text{ pF}, C_{\text{gd}} = 1 \text{ pF}$
- $R_{\text{sig}} = 10 \text{ k}\Omega, R_L = 100 \text{ k}\Omega$

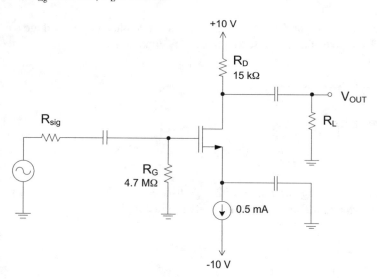

Answer It can be shown that the DC operating is $I_D = 0.5 \text{ mA}$, $V_{\text{DS}} = 5 \text{ V}$, and $V_{\text{DSAT}} = 1 \text{ V}$. Using this, the small signal parameters can be calculated.

$$g_m = 2\left(0.5\frac{\text{mA}}{\text{V}^2}\right)(1\text{V}) = 1\text{mS}$$

$$r_o = \frac{1}{\lambda \cdot I_D} = \frac{75}{0.5\,\text{mA}} = 150\,k\Omega$$

$$A_v = -\frac{4.7\,\text{M}}{10\,k+4.7\,\text{M}} \cdot (1m)(15\,k \parallel 150\,k \parallel 100\,k) = -12\frac{V}{V}$$

The equivalent input capacitance is

$$C_{in} = C_{gs} + (1-A_v) \cdot C_{gd} = 10\,\text{pF} + (1+12) \cdot 1\,\text{pF} = 23\,\text{pF}$$

Even though C_{gd} may seem small compared to C_{gs}, the Miller effect on C_{gd} makes its equivalent capacitance larger than C_{gs}. R_{sig} and C_{in} form an RC circuit with 3 dB bandwidth of

$$f_{3dB} = \frac{1}{2\pi (10\,k\Omega)(23\,\text{pF})} = 692\,\text{kHz}$$

6 Problems

1. For the RC circuit below, find the 3 dB bandwidth and calculate output voltage if the input is a 5 V peak sine wave, with a frequency equal to f_{3dB}.

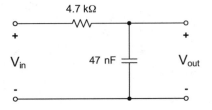

2. Assuming an ideal voltage amplifier with a gain $A_V = -10$, find the equivalent capacitance as seen at the input (Hint: the Miller effect applies here).

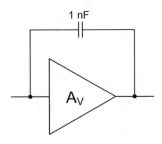

3. Suppose that the amplifier in question 2 is driven by a signal generator with an internal resistance of 1 kΩ. What is the 3 dB bandwidth of the overall circuit?

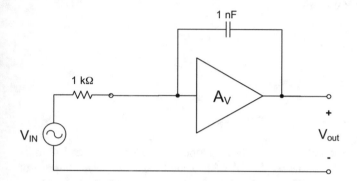

4. Find the DC operating point of the NMOS below. The parameters of the transistor are as follows:

- $K = 20$ mA/V^2
- $V_t = 2$ V
- $\lambda = 0.1$

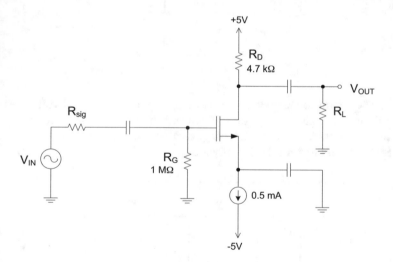

Find the midband gain and estimate the 3 dB bandwidth of this common source amplifier for each of the following cases. The NMOS transistor has $C_{gd} = 5$ pF and $C_{gs} = 60$ pF.

(a) Unloaded gain: $R_{sig} = 0$, $R_L \to \infty$
(b) Loaded gain: $R_{sig} = 10$ kΩ, $R_L = 15$ kΩ

5. Tinkercad®: Predict the 3 dB bandwidth of this circuit, and verify against simulation. Apply a 5V$_{pp}$ sine wave to the input, and adjust the frequency until the output voltage is 5/$\sqrt{2}$ V$_{pp}$.

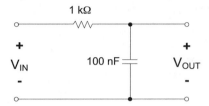

6. Tinkercad®: Find the datasheet for the 741 operational amplifier and note its pinout. Construct the following circuit with a dual polarity power supply of ±15 V. Apply a 100 Hz, 0.2Vpp sine wave to the input, and measure the output on a scope. Estimate the voltage gain of this amplifier in V/V and dB.

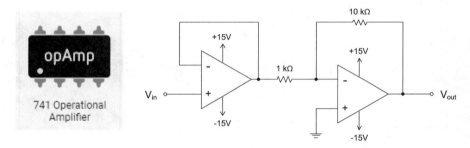

7. Tinkercad®: The circuit in the previous question approximates an ideal voltage amplifier with inverting gain. A capacitor is now added between the input and output of the amplifier. A 1 kΩ resistor is also added in series with the signal generator. Simulate and find the 3 dB bandwidth.

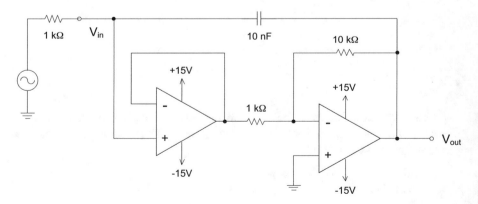

Appendix – Derivation of Common Source Amplifier Transfer Function

The common source amplifier and its frequency response are shown below. The lower cutoff f_L is set by the AC coupling capacitors, and the upper cutoff f_H is limited by the MOS capacitors C_{gs} and C_{gd}.

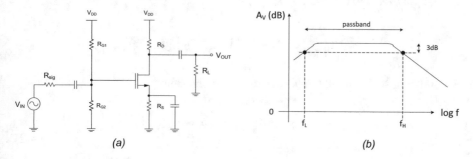

(a)   (b)

Using the NMOS high-frequency small signal model, the AC equivalent circuit is as shown.

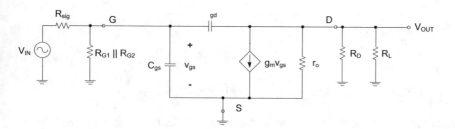

In the math below, we assume that the reader is familiar with impedances and the Laplace transform. Applying KCL to the gate node:

$$\frac{\left(v_{gs} - V_{in}\right)}{R_{sig}} + \frac{v_{gs}}{R_{G1} \parallel R_{G2}} + v_{gs} \cdot sC_{gs} + \left(v_{gs} - V_{out}\right) \cdot sC_{gd} = 0$$

Rearranging this equation such that all terms with v_{gs} are on one side:

$$v_{gs} \cdot \left(\frac{1}{R_{sig}} + \frac{1}{R_{G1} \parallel R_{G2}} + s\left(C_{gs} + C_{gd}\right)\right) = \frac{V_{in}}{R_{sig}} + V_{out} \cdot sC_{gd}$$

If we assume that $R_{G1} \parallel R_{G2} \gg R_{sig}$, we can simplify the equation to

$$v_{\text{gs}} \cdot \left(\frac{1}{R_{\text{sig}}} + s\left(C_{\text{gs}} + C_{\text{gd}}\right) \right) = \frac{V_{\text{in}}}{R_{\text{sig}}} + V_{\text{out}} \cdot sC_{\text{gd}}$$

$$v_{\text{gs}} \cdot \left(1 + s\left(C_{\text{gs}} + C_{\text{gd}}\right) \cdot R_{\text{sig}}\right) = V_{\text{in}} + V_{\text{out}} \cdot sC_{\text{gd}}R_{\text{sig}} \tag{A.1}$$

Next, apply KCL to the drain node:

$$\frac{V_{\text{out}}}{R_{\text{D}} \| r_o \| R_{\text{L}}} + g_{\text{m}}v_{\text{gs}} + \left(V_{\text{out}} - v_{\text{gs}}\right) \cdot sC_{\text{gd}} = 0$$

Rearranging such that all terms with v_{gs} are on one side:

$$v_{\text{gs}} \cdot \left(-g_{\text{m}} + sC_{\text{gd}}\right) = V_{\text{out}} \cdot \left(\frac{1}{R} + sC_{\text{gd}}\right) \tag{A.2}$$

where $R = R_{\text{D}} \| r_o \| R_{\text{L}}$. Combining Eqs. A.1 and A.2, we can find the transfer function or voltage gain of this amplifier:

$$A_{\text{v}} = \frac{V_{\text{out}}}{V_{\text{in}}} = \frac{-g_{\text{m}}R + sC_{\text{gd}}R}{a_2 \cdot s^2 + a_1 \cdot s + 1}$$

$$a_2 = C_{\text{gd}}^2 \cdot R_{\text{sig}} \cdot R$$

$$a_1 = C_{\text{gs}} \cdot R_{\text{sig}} + C_{\text{gd}} \cdot R + \left(1 + g_{\text{m}}R\right) \cdot C_{\text{gd}} \cdot R_{\text{sig}}$$

Examining the denominator, the second-order coefficient a_2 is usually small due to C_{gd}^2 and may be neglected. The first-order coefficient a_1 is the sum of multiple time constants, and note the $(1 + g_{\text{m}}R) \cdot C_{\text{gd}} \cdot R_{\text{sig}}$ is due to the Miller effect on C_{gd}, where $-g_{\text{m}}R$ is approximately the gain of the common source amplifier.

Device Physics Revisited

We will review the semiconductor physics introduced in an earlier chapter and then add more details and terminologies that will be useful as we study diodes and transistors in more depth.

1 P-Type and N-Type Silicon Review

Silicon, the most common semiconductor, has 4 valence electrons. This was represented as 4 lines shown in figure (a), where each line corresponds to a valence electron. We introduce a modified diagram in (b), where the +4 is used to indicate that overall, a silicon atom is charge neutral; the 4 valence electrons are balanced by corresponding protons. Note that the +4 does NOT refer to the atomic number of silicon (Fig. 1).

Similarly, we can represent donor and acceptor atoms as shown below (Fig. 2).

In pure silicon, the atoms arrange themselves into a regular crystalline structure, with the 4 valence electrons of one atom forming 4 covalent bonds with 4 neighboring atoms. At a temperature of absolute zero, there is no thermal energy to break the covalent bonds. As the temperature is raised, however, thermal energy allows some electrons to break loose from its bonds, in the process creating a hole as shown below; this process is called thermal *generation*. Generation creates both electrons and holes in equal numbers, referred to as electron hole pairs (EHP). Although these positive and negative charge carriers can conduct electric current, their numbers are relatively small, making pure silicon a poor conductor (Fig. 3).

While thermal generation creates EHPs, electrons and holes can collide and "annihilate" each other in a reverse process called recombination. Recombination reduces the number of EHPs. At thermal equilibrium, both generation and recombination happen continuously, resulting in a small number of EHPs for a given temperature.

C. Siu, *Electronic Devices, Circuits, and Applications*,
https://doi.org/10.1007/978-3-030-80538-8_9

Fig. 1 Stick diagram for a
silicon atom

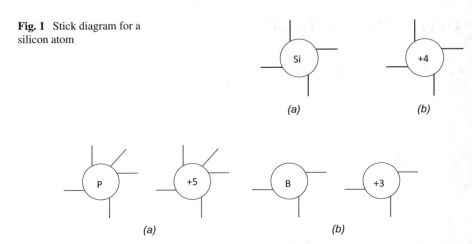

(a)   *(b)*

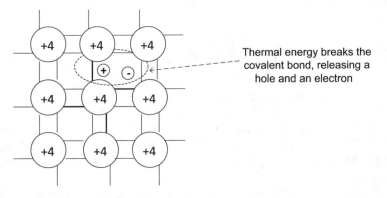

(a)   *(b)*

Fig. 2 Stick diagram for (**a**) Phosphorus donor atom (**b**) Boron acceptor atom

Fig. 3 Thermal generation in pure silicon

By introducing donor atoms (e.g., phosphorus), we can alter the conductivity of silicon. Specifically, of the 5 valence electrons for each donor atom, only 4 of them are in covalent bonds. The fifth valence electron is free to roam, increasing the number of negative carriers over that in pure silicon. Note that when the fifth valence electron wanders away from the donor atom, the donor has a net positive charge due to the missing electron. This positive charge is NOT a hole, in that the positive charge is bounded to the donor nucleus and cannot move (Fig. 4).

Similarly, by adding acceptor atoms (e.g., boron) we can also alter the conductivity of silicon. Specifically, the 3 valence electrons of each acceptor atom are in covalent bonds. Since there are four neighboring silicon atoms, one of them is missing a covalent bond. The acceptor gladly steals an electron from another covalent bond to make all its neighbors happy, but in the process creates a *hole*. We can increase the number of positive charge carriers over pure silicon by adding acceptor atoms (Fig. 5).

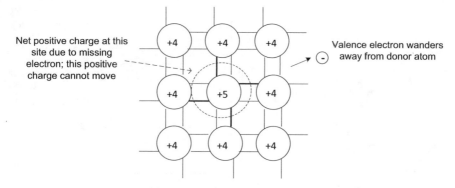

Net positive charge at this site due to missing electron; this positive charge cannot move

Valence electron wanders away from donor atom

Fig. 4 N-type silicon with electrons as majority carriers

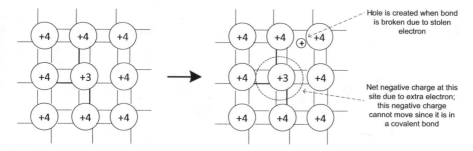

Hole is created when bond is broken due to stolen electron

Net negative charge at this site due to extra electron; this negative charge cannot move since it is in a covalent bond

Fig. 5 P-type silicon with holes as majority carriers

In summary,

- By adding donor atoms, n-type silicon has *majority carriers* that are negative (electrons).
- By adding acceptor atoms, p-type silicon has *majority carriers* that are positive (holes).

2 Energy Band Diagrams

In introductory quantum mechanics, we are told that electrons can only have discrete energy levels in an atom. Shown below are the electron orbitals or energy levels for a silicon atom. The inner orbits ($n = 1, 2$) are filled with electrons. The valence orbit with $n = 3$ has 4 electrons, with 2 in the 3s orbital and 2 in the 3p orbital. The 3s and 3p orbitals can hold a total of 8 electrons, and thus we can say that the valence orbit has 8 possible states, and only half are filled with the 4 valence electrons (Fig. 6).

If there are N isolated silicon atoms, we can expect that there will be 8 N possible states in the valence orbits. If the N silicon atoms are brought together to form a

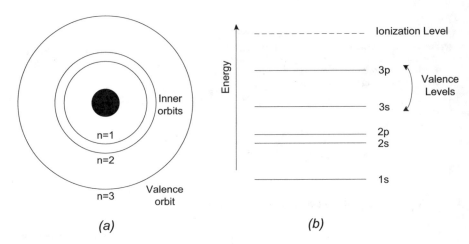

Fig. 6 Single silicon atom (**a**) electron orbitals (**b**) electron energy levels

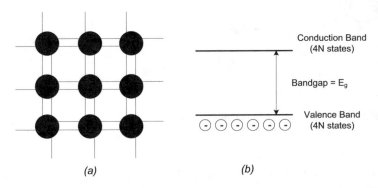

Fig. 7 Silicon crystal (**a**) silicon atoms in covalent bonds (**b**) silicon crystal: electron energy levels

crystal, these states cannot remain the same due to the Pauli exclusion principle. In a process called *sp3 hybridization*, the state cluster is split into two groups: one group is called the valence band with 4N states, and the other is called the conduction band with 4N states also. In the valence band, for example, the 4N states are not identical, but split into 4N closely spaced energy levels.

As shown in the diagram below, the valence band and the conduction band are separated by an energy known as the bandgap or E_g. This is the energy that an electron must receive to jump from the valence to conduction band. At a temperature of absolute zero, there is no thermal energy, and valence band's 4N states are completely filled with the 4N valence electrons. The conduction band is devoid of electrons, and hence pure silicon at 0K cannot conduct electrical current (Fig. 7).

As the temperature is raised, some electrons in the valence band will have sufficient thermal energy to jump to the conduction band. Electrons in the conduction band can carry electric current. We can quantify the energy required to break a covalent bond, which is equal to the bandgap E_g. In silicon, E_g is **1.1** electron volt

Fig. 8 Thermal energy breaking covalent bond, shown with energy band diagram

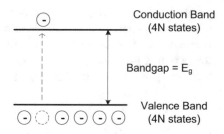

(eV), where 1 eV is defined as the amount of energy required to raise the potential of a single electron by 1 volt (1 eV = 1.6×10^{-19} Joules) (Fig. 8).

As shown in the diagram, when one electron jumps to the conduction band, it leaves behind a *hole* in the *valence band*. Hence in pure silicon, also referred to as an *intrinsic* silicon, the number of conduction band electrons is equal to the number of valence band holes, resulting in electron-hole pairs (EHP). Mathematically, this can be expressed as

$$n = p = n_i$$

- n = electron concentration in the conduction band per cm^3,
- p = hole concentration in the valence band per cm^3,
- n_i = intrinsic carrier concentration, equal to 0 at absolute zero, but increases as temperature goes up.

Using the above equation, we can derive an expression regarding the product of n and p:

$$n \bullet p = n_i^2$$

The product of the electron and hole concentrations is equal to n_i^2, which we can regard as constant for a given temperature. This expression has general applicability that will be explored later.

In summary:

- Within a silicon crystal, there are 5×10^{22} silicon atoms per cubic centimeter.
- At absolute zero, there are no electrons in the conduction band; all electrons are in the valence band.
- As the temperature is raised, some electrons will gain sufficient energy to jump to the conduction band, also creating holes in the valence bands; the creation of electron hole pairs (EHP) can be visualized using an energy band diagram.
- The energy required to jump from the valence to the conduction band is called the bandgap energy, which is **1.1 electron Volt** (eV) for silicon.
- At room temperature, there are approximately **1.5×10^{10}** electrons/cm^3 in the conduction band; this is referred to as the intrinsic carrier concentration n_i.

- While 1×10^{10} is a large number, note that this is a tiny fraction of the 5×10^{22} silicon atoms per cm^3. To put this into perspective, a beach that is the size of a football field is estimated to have roughly 10 trillion grains of sand. At room temperature, the number of electrons in the conduction band corresponds to less than one grain of sand on this beach!

Energy Band Diagram for N-Type Silicon

Pure silicon is a poor conductor even at room temperature. By introducing column V elements into the silicon crystal, we can improve its conductivity in a process known as doping. Doped silicon is also referred to as *extrinsic* silicon, and we can view the donor atoms as the addition of a new energy level E_d in the energy band diagram. The level E_d is much closer to the conduction band, and hence most donor atoms will be ionized at room temperature; the fifth valence electrons from most donor atoms will be in the conduction band (Fig. 9).

As a result, the electron concentration in the conduction band is greatly increased, improving the conductivity of n-type silicon via these negative charge carriers. Let the donor atom concentration be N_D atoms/cm^3; assuming that all donor atoms are ionized and that $N_D \gg n_i$, then the electron concentration is approximately N_D. Previously, we had noted that $n{\cdot}p = n_i^2$. With $n \sim N_D$, the hole concentration is

$$p = \frac{n_i^2}{N_D}$$

Example A silicon wafer is doped with phosphorus atoms to a concentration $N_D = 1 \times 10^{16}$ atoms/cm^3. At room temperature, what are the hole and electron concentrations?

Answer At room temperature, all the phosphorus atoms will be ionized, meaning that the electron concentration is $N_D + n_i$. Since $N_D \gg n_i$, it is common to approximate the electron concentration as equal to N_D.

$$n = n_o \sim N_D = 1 \times 10^{16} \text{ electrons / cm}^3$$

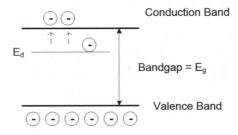

Fig. 9 Energy band diagram of N-type silicon

We can then calculate the hole concentration by

$$p = p_o = \frac{n_i^2}{N_D} = \frac{\left(1.5 \times 10^{10}\right)^2}{1 \times 10^{16}} = 22500 \, \text{holes} \, / \, \text{cm}^3$$

Electrons are the *majority carriers* in n-type silicon, as seen in how they vastly out-number the holes. Holes are also called minority carriers in n-type silicon.

Energy Band Diagram for P-Type Silicon

By putting acceptor atoms into pure silicon, we can increase the number of holes and improve its conductivity. We can view the acceptor atoms as the addition of a new energy level E_a in the energy band diagram. The level E_a is near the valence band, and hence electrons in the valence band can easily jump up to E_a. At room temperature, nearly all the atoms or states at E_a are filled with electrons, creating an equal number of *holes* in the valence band (Fig. 10).

As a result, the hole concentration in the valence band is greatly increased, and electric current can be conducted by these positive charge carriers. Let the acceptor atom concentration be N_A atoms/cm³; assuming that all acceptor atoms are ionized and that $N_A \gg n_i$, then the hole concentration is approximately N_A. Previously, we had noted that $n \cdot p = n_i^2$. With $p \sim N_A$, the electron concentration is

$$n = \frac{n_i^2}{N_A}$$

Example A silicon wafer is doped with boron atoms to a concentration $N_A = 2 \times 10^{15}$ atoms/cm³. At room temperature, what are the hole and electron concentrations?

Fig. 10 Energy band fiagram of P-type silicon

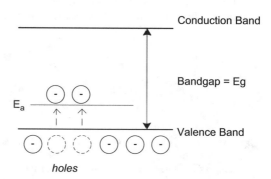

Answer At room temperature, all the boron atoms will be ionized, meaning that the hole concentration is $N_A + n_i$. Since $N_A \gg n_i$, it is common to approximate the hole concentration as equal to N_A.

$$p = p_o \sim N_A = 2 \times 10^{15} \text{ holes / cm}^3$$

We can then calculate the electron concentration by

$$n = n_o = \frac{n_i^2}{N_A} = \frac{\left(1.5 \times 10^{10}\right)^2}{2 \times 10^{15}} = 112500 \text{ electrons / cm}^3$$

Holes are the majority carriers in p-type silicon, as seen in how they vastly outnumber the electrons. Electrons are called minority carriers in p-type silicon.

3   Diffusion and Drift Currents

Having established carrier concentration in both intrinsic and doped silicon, we can now study the movement of these carriers that will give rise to electrical current. There are two processes involved: diffusion and drift.

Diffusion Current

Diffusion of carriers is similar conceptually to the diffusion of gas or liquid: carriers move from an area of high concentration to that of a lower concentration, and this in turn creates current flow. As seen in the figure, (a) shows a situation where there is a barrier preventing the electrons from going to the empty right hand side. In (b), the barrier is lifted, and electrons diffuse to the right, resulting in a *diffusion current* (Fig. 11).

Intuitively, we would expect the concentration gradient to affect the magnitude of the diffusion current, and this can be expressed mathematically as.

$$J_n = qD_n \frac{dn}{dx} \quad \text{for electrons}$$

$$J_p = -qD_p \frac{dp}{dx} \quad \text{for holes}$$

where

- J_n, J_p = diffusion current density (A/cm^2).

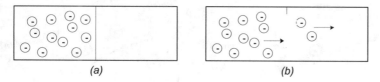

Fig. 11 Electron diffusion

- D_n, D_p = diffusion coefficient for electrons and holes respectively (cm²/s).
- dn/dx, dp/dx = electron and hole concentration gradients (# of carriers /cm⁴),
- q = electronic charge = 1.6×10^{-19} C (Coulombs).

The diffusion coefficient quantifies the ease of carrier diffusion in response to a carrier gradient, noting that the carriers are travelling through a solid and will collide with atoms in the crystal.

The signs for J_n and J_p are different due to the definition of conventional current flow and are worth examining here:

- Conventional current flow = direction of positive charge movement.
- Diffusion current due to holes is the same direction as hole movement; use positive sign.
- Diffusion current due to electrons is the opposite direction as the electron movement; use negative sign.
- Carrier movement = charges will diffuse from areas of high concentration to lower concentration; in other words, the carriers will move in the direction where the gradient (e.g., dn/dx) is negative.
- Note then $J_n = q \cdot D_n \cdot dn/dx$ is negative with a negative gradient dn/dx, while $J_p = -q \cdot D_p \cdot dp/dx$ is positive with a negative gradient dp/dx.

These conventions and definitions are summarized in the figure (Fig. 12).

The total diffusion current due to electrons and holes moving inside a solid is thus equal to

$$J_{\text{diff}} = qD_n \frac{dn}{dx} - qD_p \frac{dp}{dx}$$

Drift Current

Aside from diffusion, charge carriers will also move in the presence of an electric field. Just like diffusion, the carriers will collide with other atoms as they travel through the silicon crystal, and so the carrier velocity is influenced by these collisions. In semiconductor physics, the drift velocity is characterized by parameters called electron mobility and hole mobility.

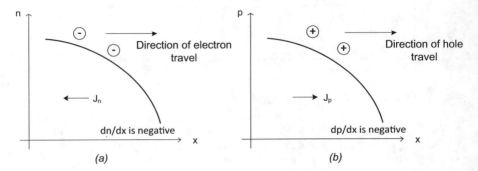

Fig. 12 Diffusion current for (**a**) electrons, (**b**) holes

$$\mu_n = \text{Electron mobility} \left(cm^2 / (V{\cdot}s) \right)$$

$$\mu_p = \text{Hole mobility} \left(cm^2 / (V{\cdot}s) \right).$$

The mobility parameters for silicon depend on temperature and the doping concentration, but typical values are $\mu_n = 1200 \ cm^2/(V{\cdot}s)$ and $\mu_p = 500 \ cm^2/(V{\cdot}s)$. From this, we see that holes travel slower than electrons, or holes appear "heavier" than electrons. An intuitive explanation is that electrons in the conduction band can move much easier than holes in the valence band. While one can argue that hole motion in the valence band ultimately involves electron movement, the difference is that we need to break a covalent bond and steal an electron for holes to move, and this is a more cumbersome process.

Next, as we apply an electric field to p-type or n-type silicon, the carriers will move faster for a stronger field. This trend continues until the carrier collision imposes a speed limit, where a stronger field will not increase the speed further, in a process known as velocity saturation. For fields and speeds under the limit, the drift current densities are given by.

$$J_n = qn \cdot \mu_n E \qquad \text{for electrons}$$

$$J_p = qp \cdot \mu_p E \qquad \text{for holes}$$

In the above, n and p stands for the electron and hole carrier concentrations, respectively. The signs for J_n and J_p are explained using the diagram below. By convention, holes travel in the same direction as the electric field, while electrons travel in the opposite direction. When conventional current flow is applied, both the electron current and hole current travel in the same direction as the electric field; the total drift current is the sum of these currents (Fig. 13):

$$J_{\text{drift}} = qn \bullet \mu_n E + qp \bullet \mu_p E$$

In summary, diffusion current is caused by concentration gradient in the semiconductor, while drift current occurs due to the application of an external electric field. Both types of current can flow at the same time depending on the semiconductor device that is used.

4   The Physics of P-N Junctions

N-type silicon has majority carriers that are negative, but since the donor atoms have released its fifth valence electrons, there is a net positive charge associated with the donor that cannot move. We can depict a majority carrier as a negative sign with a bubble around it; the bubble indicates that this charge can move around. On the other hand, the positive donor sites are shown as plus signs without a bubble, indicating that they are immobile. Note that overall, the n-type silicon is charge neutral (Fig. 14).

We can make a similar diagram for p-type silicon, where the majority carriers are positive and shown as positive signs with bubbles. Each acceptor atom, having stolen an electron to complete the fourth covalent bond, has a net negative charge that cannot move (Fig. 15).

When we join a piece of p-type silicon with n-type silicon, diffusion occurs due to the different carrier concentrations. Described another way, since the number of holes is high in p-type silicon and low in n-type silicon, the holes move from p-type to n-type via diffusion. The same process happens with the electrons in n-type silicon, diffusing over to p-type silicon.

During diffusion, holes from the p-type silicon can collide with the electrons from n-type silicon, and recombination will eliminate these charge carriers. Near the boundary between p-type and n-type silicon, recombination will eliminate free holes and electrons, leaving behind immobile acceptor and donor sites. This region that is depleted of charge carriers is called the *depletion region* (Fig. 16).

Does the diffusion continue until recombination eliminates all charge carriers? The answer is *no*, as donor and acceptor sites in the depletion region create an electric field. The electric field prevents further diffusion from happening. We can visualize the depletion region as a wall or potential barrier that obstructs the further diffusion of charge carriers (Fig. 17).

Another valid way to view this electric field is to look at the associated drift current. Recall that holes and electrons will move in response to an electric field. The P-N junction's built-in electric field will create electron and hole drift currents, while the concentration gradient results in electron and hole diffusion currents. In thermal equilibrium, without any external voltage applied to the P-N junction, the drift and diffusion currents must *cancel* each other (Fig. 18).

In the figure above, note that while the current in the external circuit $i = 0$, drift and diffusion currents are flowing simultaneously inside the junction. Since the drift and diffusion components must cancel in this case, we have

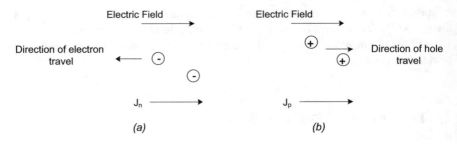

Fig. 13 Drift current for (**a**) electrons, (**b**) holes

Fig. 14 N-type silicon
charge diagram

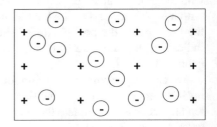

Fig. 15 P-type silicon
charge diagram

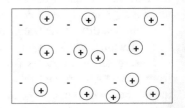

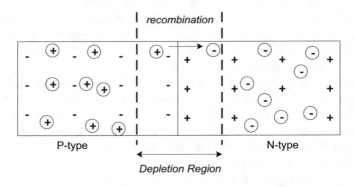

Fig. 16 P-N junction charge diagram

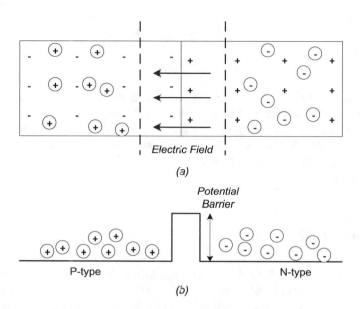

Fig. 17 Electric field (**a**) inside P-N junction, creating a potential barrier (**b**)

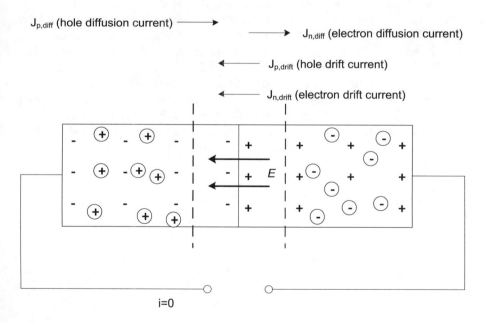

Fig. 18 Currents inside a P-N Junction with no external bias

$$J_{n,\text{diff}} + J_{n,\text{drift}} = 0$$

$$J_{p,\text{diff}} + J_{p,\text{drift}} = 0$$

Let us make one more observation on the drift currents before we move on. For example, the hole drift current consists of holes flowing from the n-type silicon to the p-type silicon. However, holes are minority carriers in n-type silicon, in that electrons vastly outnumber holes in n-type silicon. This gives us some intuition that *drift* is a limited mechanism in P-N junctions, and we will explore this further later in the chapter.

The Fermi Level

Quantum mechanics is the basis for semiconductor physics, and the present interpretation of quantum physics uses probability theory. The distribution of electrons over a range of allowed energy levels obeys *Fermi-Dirac* statistics. While we will not dive into the detailed mathematics here, one key result from Fermi-Dirac statistics is the concept of the *Fermi Level E_F* in semiconductors. For now, we will simply state the Fermi levels for intrinsic and extrinsic silicon without proof (Fig. 19).

For intrinsic silicon, the Fermi level is at the middle of the bandgap. For n-type silicon, E_F is closer to the conduction band, while for p-type silicon, E_F is closer to the valence band. These results are valid at equilibrium, and with no external bias, *the Fermi level for different silicon types must all line up to form a horizontal line.* This idea is illustrated in the figure below for a P-N junction (Fig. 20).

As a result of E_F forming a horizontal line, note that both the conduction and valence bands are bent in the energy band diagram. This band bending is a representation of the potential barrier mentioned earlier. V_o is the built-in potential of a P-N junction, which we have assumed to be 0.7 V in our previous work.

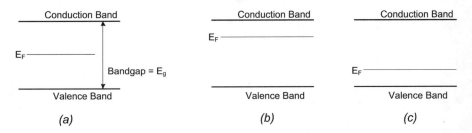

Fig. 19 Fermi level for (**a**) intrinsic silicon (**b**) n-type silicon (**c**) p-type silicon

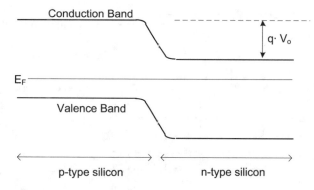

Fig. 20 Energy band diagram for P-N junction

P-N Junction Under Forward Bias

If we apply an external voltage V_f to the diode in the forward direction, we know from the simple diode model that the current can increase rapidly with a voltage of 0.7 V across the diode. We can now present a more detailed model for the diode, in that the diode current I_D increases exponentially with the diode voltage V_D:

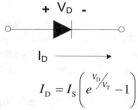

$$I_D = I_S \left(e^{V_D / V_T} - 1 \right)$$

- I_S = saturation current; a device parameter dependent on the geometry, doping level, etc. of the diode.
- V_T = thermal voltage, not to be confused with the MOSFET threshold voltage.
- $V_T = kT/q$, where k = Boltzmann's constant, T = temperature in Kelvin, and q = electron charge. A commonly used value for V_T is 26 mV, which is approximately the thermal voltage at $T = 300$ K.

While we will not derive this equation, we can gain some understanding by looking at the device physics. Shown below is a P-N junction under forward bias. One way to understand the diode action is to see that the "+" terminal of the battery is repelling the holes in the p-type, while the battery "–" terminal is repelling the electrons in the n-type. In this way, we can visualize the battery V_f as acting to increase diffusion: holes are pushed from the p-type to the n-type material, increasing the hole diffusion current. A similar argument can be made for electrons in the n-type silicon (Fig. 21).

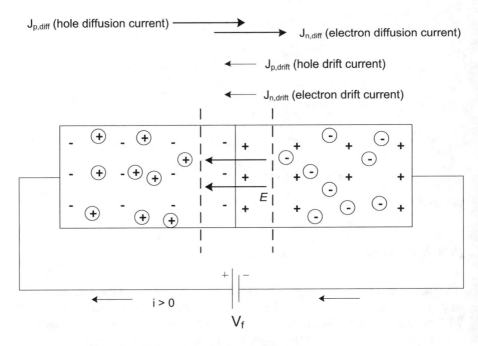

Fig. 21 Currents inside a P-N Junction with forward bias

For reasons that we will not explore in detail, the drift current does not change much going from zero to forward bias. One way to view this is that drift is a limited mechanism in the diode as explained earlier. As a result, *diffusion current* is much bigger than the drift current for a diode in forward bias, as indicated by the relative sizes of the arrows in the figure.

Another way of studying a diode in forward bias is using an energy band diagram. Repeated below is the band diagram of a diode with no external bias, except carriers are shown now. In the n-type, the majority carriers are electrons in the conduction band. To move to the p-type side, however, these electrons need to climb "uphill"; this climb is the potential barrier that was mentioned earlier (Fig. 22).

Conversely in p-type silicon, holes in the valence band are the majority carriers. To cross over to the n-type, holes need to climb "downhill" (this is opposite to intuition, but potential barriers for holes are downhill slopes).

When an external voltage V_f is applied, the rule regarding Fermi level changes: the Fermi level separates between the p-type and n-type, reducing the potential barrier. The energy band diagram will therefore change to the following (Fig. 23):

For example, electrons in the n-type silicon now have a smaller uphill climb, and the number of electrons that can travel over to the p-type increases exponentially. Both the hole and electron *diffusion currents increase exponentially* with applied voltage, allowing a forward-biased diode to conduct electric current.

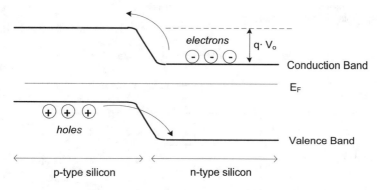

Fig. 22 Energy band diagram for P-N Junction with no external bias

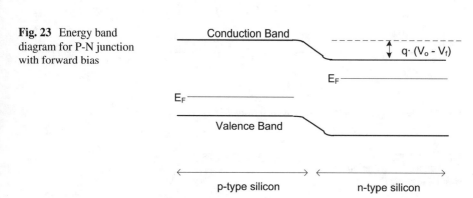

Fig. 23 Energy band diagram for P-N junction with forward bias

P-N Junction Under Reverse Bias

For a P-N junction in reverse bias V_r, note that the "+" terminal of the battery is now connected to the diode's cathode, which is n-type silicon. We can visualize that the battery "+" terminal is attracting the majority carriers in the n-type silicon, reducing the electron diffusion. Similarly, the battery "-" terminal is inhibiting the hole diffusion in p-type silicon. As a result, the diffusion currents are drastically reduced in reverse bias, with only the drift currents flowing in the diode (Fig. 24).

As the relative sizes of the arrows show, diffusion currents are reduced to almost nothing in reverse bias. The drift currents, being a limited mechanism in a P-N junction, are small but still flow through the diode. Hence for a diode in reverse diode, the current is not exactly zero as in the simple model, but is very small, typically in the nA range.

Using an energy band diagram, we can see that the reverse bias V_r increases the potential barrier. Carriers cannot climb this barrier, making the diffusion currents disappear (Fig. 25).

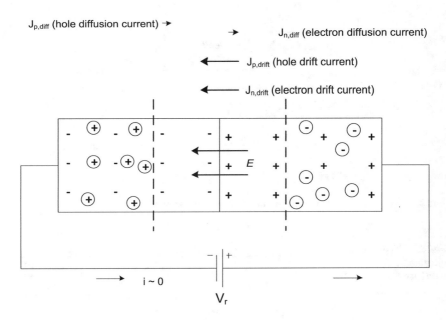

Fig. 24 Currents inside a P-N junction with reverse bias

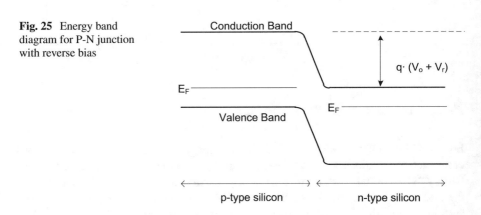

Fig. 25 Energy band diagram for P-N junction with reverse bias

In summary:

- For a P-N junction in equilibrium with no applied bias, the diffusion and drift currents cancel each other, resulting in a net current of zero through the diode.
- A P-N junction contains a depletion region with no charge carriers; this depletion region generates an electric field, which is a potential barrier for carrier diffusion.
- By applying an external forward bias voltage, the potential barrier is reduced, allowing the diffusion currents to increase exponentially.
- In a forward-biased diode, the diode current is dominated by the diffusion process.

- In a reverse-biased diode, the diffusion currents are reduced to nothing, leaving only a small drift current flowing in the reverse direction.

5 Problems

1. In pure silicon at room temperature, in the absence of any applied voltage, what are the physical processes that determine the number of electron-hole pairs?
2. What are the majority carriers in p-type silicon?
3. What are the majority carriers in n-type silicon?
4. In an energy band diagram of intrinsic silicon, what is the separation between the valence band and conduction band called?
5. A silicon wafer is doped with phosphorus atoms to a concentration $N_D = 5 \times 10^{16}$ atoms/cm^3. At room temperature, what are the hole and electron concentrations?
6. In a silicon diode, what happens at the boundary between p-type silicon and n-type silicon?
7. In the energy band diagram for intrinsic silicon, where is the Fermi Level situated? How does the Fermi Level change with doping, for n-type silicon and p-type silicon?
8. For a silicon diode in forward bias, what is the main mechanism for carrier transport/current flow?
9. For a silicon diode in reverse bias, why is there a small current that flows? What is the mechanism for this current?

Diode Circuits

We had studied that diodes are used to build rectifier circuits and perform the important task of AC to DC conversion. In this chapter, we will discuss other diode circuits that alter AC waveforms, special diodes such as Schottky diode and Zener diode, and voltage regulators.

1 Clipping Circuits

Clipping circuits are used to remove parts of a waveform. We saw this action earlier with the half wave rectifier, which discarded half of input signal. We had used an analysis procedure where we found V_{OUT} when the diode is off and then deduced whether the diode is on or off as V_{IN} cycled through different values. This procedure can still be used to analyze the clipping circuits, but we will introduce a systematic technique specific to these circuits:

1. Determine V_{out} when the diode is on.
2. Determine V_{out} when the diode is off.
3. Examine the range of V_{in} for which the diode is on and the range for which it is off. In conjunction with the information from the first two steps, we can sketch the output waveform.

Series Clipping Circuits

If the diode is in series between the input and output, then we have a series clipping circuit as shown below. We had analyzed this circuit in an earlier chapter, in which it was called a half wave rectifier (Fig. 1).

Following the procedure, we answer the three questions:

© Springer Nature Switzerland AG 2022
C. Siu, *Electronic Devices, Circuits, and Applications*,
https://doi.org/10.1007/978-3-030-80538-8_10

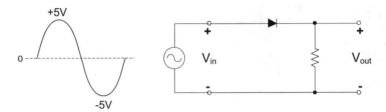

Fig. 1 Series clipping circuit

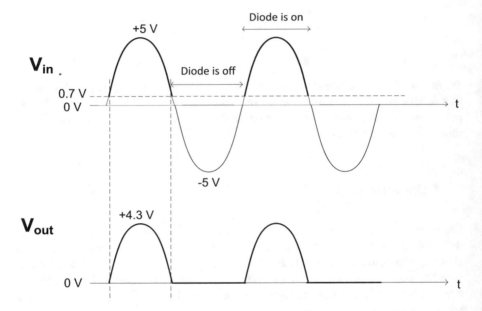

Fig. 2 Operation of the series clipping circuit

- When the diode is on, $V_{OUT} = V_{IN} - 0.7$ V.
- When the diode is off, $V_{OUT} = 0$ V.
- The diode is on for $V_{IN} \geq 0.7$ V.

We then combine this information together to figure out what the output looks like. One way is to annotate the input waveform, enabling us to visualize the output:

- On the input waveform, highlight the time intervals in which the diode is on.
- During those intervals, $V_{OUT} = V_{IN} - 0.7$ V, and hence the output follows the input but with a 0.7 V drop.
- In the other time intervals, the diode is off and $V_{OUT} = 0$ V (Fig. 2).

The output waveform coincides with the half wave rectifier results, of course. In this circuit, we have clipped roughly half of the input. What if we wish to cut not half, but some other portion? We can introduce a battery or voltage source in the clipping circuit for this purpose.

Example Find the output waveform for the circuit below, for a +5 V peak sine wave input:

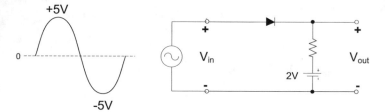

Answer Following the procedure and answering the three questions:

- When the diode is on, $V_{OUT} = V_{IN} - 0.7$ V.
- When the diode is off, $V_{OUT} = +2$ V.
- The diode is on for $V_{IN} \geq 2.7$ V.

Annotating this information onto the input waveform provides us with a sketch of the output.

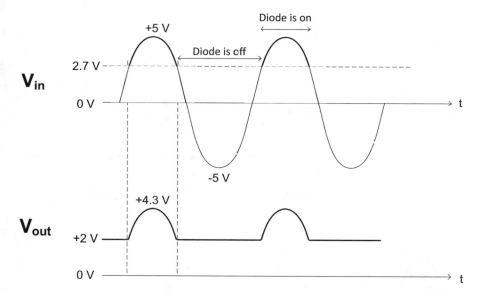

In this case, less than half of the input is retained, with a 2 V floor when the diode is off.

Shunt Clipping Circuits

If the diode is in shunt across the output, then we have a shunt clipping circuit as shown below. Following the procedure and answering the three questions:

- When the diode is on, $V_{OUT} = 0.7$ V.
- When the diode is off, $V_{OUT} = V_{IN.}$
- The diode is on for $V_{IN} \geq 0.7$ V (Fig. 3).

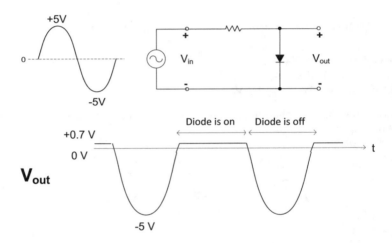

Fig. 3 Operation of the shunt clipping circuit

Note that the shunt clipper retains half of the input without the 0.7 V drop, due to the fact that the diode is off when V_{OUT} follows V_{IN}.

Example Find the output waveform for the circuit below, for a + 5 V peak sine wave input:

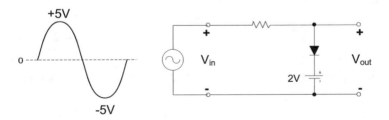

Answer this is a shunt clipper with battery bias, allowing us to adjust how much of the input to keep. The answers to the three questions:

- When the diode is on, $V_{OUT} = 2$ V $+ 0.7$ V $= 2.7$ V.
- When the diode is off, $V_{OUT} = V_{IN}$.
- The diode is on for $V_{IN} \geq 2.7$ V.

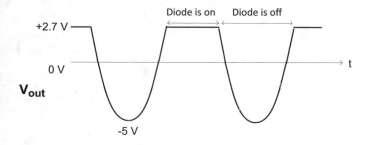

Applications of Clipping Circuits

One application of clippers is electrostatic discharge (ESD) protection for integrated circuits. A common form of ESD we have all experienced is the shock received from touching a door knob on a dry winter day. By walking around on a carpet, we build up static electricity on our body upwards of several thousand volts. If we touch an integrated circuit, the circuit can be damaged by the ESD. The sudden high voltage discharge can rupture the gate oxide of MOSFETs, permanently damaging those devices (Fig. 4).

In almost all modern ICs, ESD protection is included on the chip, meeting standards such as the human body model (HBM) at 2000 Volts. The protection can be as simple as a resistor and two diodes, connected in a shunt configuration as shown below. The resistor R limits the current during the ESD strike, with the diodes and power clamp limiting the voltage on the NMOS gate (Fig. 5).

When ESD strikes the input of an IC, the voltage can be positive or negative with respect to ground, and the protection circuit must handle both cases. When the ESD voltage is positive, D1 and the power clamp turn on. A power clamp is a protection circuit that is off for normal V_{DD} values. However, if V_{DD} rises beyond a certain limit, the power clamp turns on and fixes the voltage between V_{DD} and ground to V_{clamp}. In

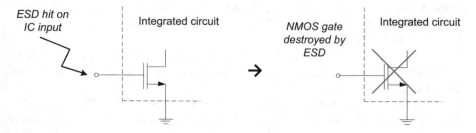

Fig. 4 Electrostatic Discharge (ESD) and its effects

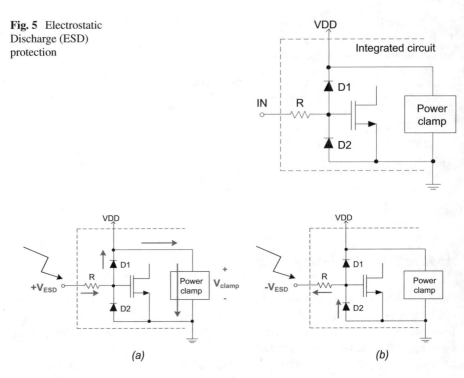

Fig. 5 Electrostatic Discharge (ESD) protection

Fig. 6 ESD protection (**a**) during a positive hit (**b**) during a negative hit

this way, the voltage on the NMOS gate is limited to $V_{clamp} + 0.7$ V, which will be a safe value to ensure the NMOS transistor survives (Fig. 6).

When the ESD voltage is negative, the diode D2 turns on and limits the gate voltage to -0.7 V. In this way, we see that the IC input is protected against positive and negative ESD events.

2   Clamping Circuits

There are circuits that clamp the peak of a waveform to a specified voltage. For example, suppose that we wish to clamp the positive peak of the following input to 0 V. Ideally, the output is still a 5 V peak sine wave, but the average value is no longer zero (Fig. 7).

We recognize that the output is the same as the input but with some DC offset added. One possible implementation is to put a battery in series with the signal generator to create this shift (Fig. 8):

The issue with this implementation is that it only works for a 5 V peak waveform. If the input changed to 2 V peak sine wave, then the battery voltage will also need to change to keep the positive peak clamped at 0 V. Is there a way to

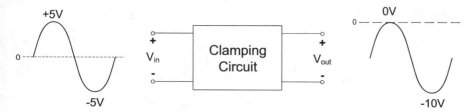

Fig. 7 Clamping circuit

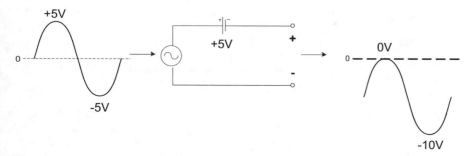

Fig. 8 Clamping using a series battery

automatically set the correct battery voltage? The answer lies in replacing the battery with a capacitor and charging that capacitor to the right voltage. A capacitor will maintain the same voltage as long as the discharge is small, or it is recharged periodically.

A circuit that can automatically charge a capacitor to the correct voltage is shown below, comprised of a diode and a capacitor. We will analyze this circuit in two different ways. The first method involves tracking how the diode and capacitor behave as the input varies; if you understand the first method, you will have a core understanding of this circuit's operation. The second method is more procedural and helps us to analyze this type of circuit faster (Fig. 9).

Clamping Circuit: Analysis Method #1

In the first method, we begin by assuming the capacitor is not charged ($v_C = 0$), and finding how the input waveform can turn on the diode and charge the capacitor. We will assume a sine wave is applied to the input:

First Quarter Cycle of Input (0 to 90 Degrees)

- Assume that at $t = 0$, the capacitor is not charged ($v_c = 0$) and $V_{in} = 0$. Hence $V_{out} = 0$ and the diode is off.

Fig. 9 Simple clamping circuit

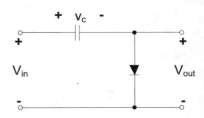

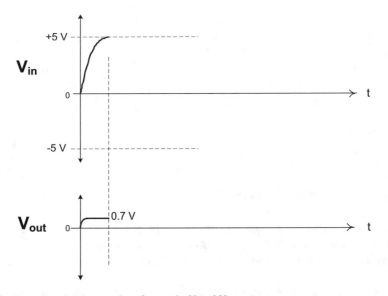

Fig. 10 Clamping circuit operation, first cycle 0° to 90°

- Since the diode is off, it is an open circuit, and the capacitor has no path for charging or discharging. As V_{in} rises from 0 V, V_{out} follows V_{in}, until V_{in} reaches 0.7 V. At that point in time, the diode turns on, and the V_{out} gets clamped to 0.7 V.
- Once the diode is on, the capacitor has a path for current flow. As V_{in} rises above 0.7 V, the capacitor charges, and v_c increases.
- As the input rises to its peak of +5 V in the first quarter cycle, the capacitor charges to $v_c = +5 - 0.7 = 4.3$ V (Fig. 10).

Second Quarter Cycle of Input (90 to 180 Degrees)
- At the end of the first quarter cycle, the capacitor voltage is $v_c = 4.3$ V.
- The capacitor acts like a battery in series with the input V_{in}, so $V_{out} = V_{in} - 4.3$ V.
- The input V_{in} starts to decrease in voltage, but with $v_c = 4.3$ V, V_{out} goes below 0.7 V. The diode turns off, and the capacitor once again has no path for charging/discharging. v_c remains at 4.3 V.
- The diode is off for the second quarter cycle, and when V_{in} reaches 0 V, $V_{out} = 0 - 4.3 = -4.3$ V (Fig. 11).

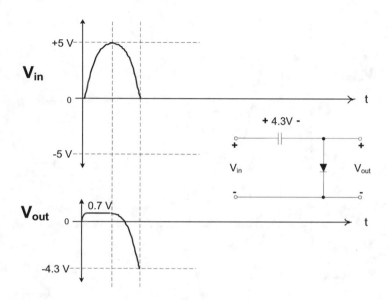

Fig. 11 Clamping circuit operation, first cycle 90° to 180°

Third and Fourth Quarter Cycles of Input (180 to 360 Degrees)
- V_{in} is negative for the remainder of the sine wave period
- V_{out} continues to follow V_{in}. The diode remains off since V_{out} is negative (diode is reverse biased)
- The capacitor acts like a 4.3 V battery in series with V_{in}, hence $V_{out} = V_{in} - 4.3$ V
- V_{in} reaches the negative peak of −5 V, $V_{out} = -5 - 4.3 = -9.3$ V (Fig. 12)

After the First Cycle
- For an ideal diode with no reverse leakage current, the capacitor keeps its voltage of 4.3 V.
- The reason why the capacitor holds its charge is that the diode remains off in subsequent cycles.
- At the instant when Vin rises back to its peak of +5 V, the output V_{out} goes to a peak of +0.7 V. With $V_{out} = 0.7$ V, the diode turns on briefly, but since the potential difference between V_{in} and V_{out} is 4.3 V, there is no reason for the capacitor voltage to change. v_c remains at 4.3 V.
- After the initial startup, the output is a 5 V peak sine wave with the positive peak clamped to 0.7 V, as seen in the figure below (Fig. 13).

In summary, analysis method #1 shows us how to start from zero initial conditions and find the final capacitor voltage. The capacitor then acts like a battery that adds a DC offset to the input waveform, effectively clamping one of the peaks to a desired voltage.

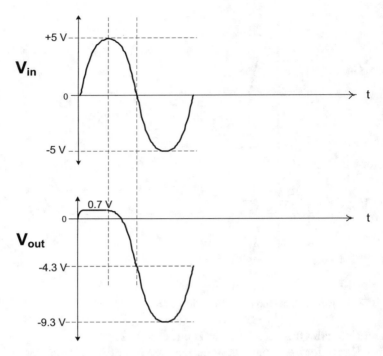

Fig. 12 Clamping circuit operation, first cycle 180° to 360°

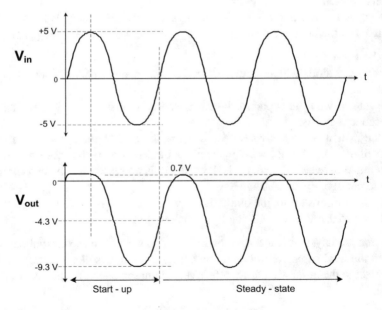

Fig. 13 Clamping circuit operation, steady state

Exercise For a 5 V peak sine wave input, sketch the output of the following circuit in steady state. Also determine the capacitor voltage in steady state.

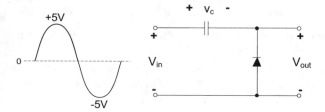

Answer Compared to the previous circuit, the diode direction is reversed in this circuit. Starting from zero initial conditions, we find that the diode turns on when the input goes negative in the first cycle. As a result, the capacitor voltage $v_c = -5$ V $- (-0.7$ V$) = -4.3$ V.

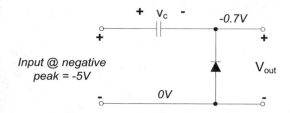

With $v_c = -4.3$ V, by KVL we will find that the input is shifted up by this amount. In other words, the negative peak of the input is clamped to -0.7 V.

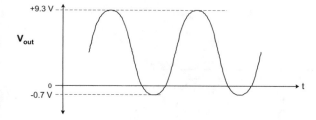

In the clamping circuits presented so far, we can clamp the positive peak or negative peak of the input based on the diode direction. However, note that we are clamping at either $+0.7$ V or -0.7 V due to the diode drop. Can we clamp to a different voltage? Just as in the clipping circuits, we can add a battery, and this is also a good time to introduce analysis method #2.

Clamping Circuit: Analysis Method #2

In the second method, we answer some simple questions to determine the output waveform. The method is illustrated by analyzing the following circuit (Fig. 14):

Step 1: Calculate the *clamping voltage*, which is the value of V_{OUT} when the diode is on. For this circuit, the clamping voltage is −0.3 V.

Step 2: Based on the diode direction, identify whether the positive peak or negative peak is clamped. In this circuit, the positive peak is clamped based on the earlier examples that we have done.

If we can perform step 1 and 2 correctly, then we can sketch the output as a 5 V peak sine wave (10 V peak to peak), with the positive peak clamped to −0.3 V.

Step 3: Determine the steady-state voltage across the capacitor, and check if the result from step 2 is correct. Since the positive peak is clamped, at the positive peak the input is +5 V and the output is at −0.3 V; hence, v_c = 5 V − (−0.3 V) = 5.3 V. We can verify that a capacitor charged to +5.3 V does indeed shift the input waveform as shown above (Fig. 15).

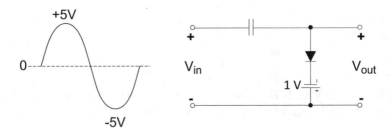

Fig. 14 Biased clamping circuit

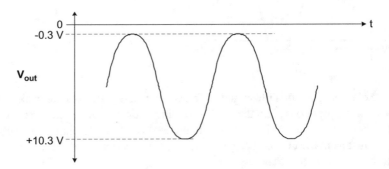

Fig. 15 Output of biased clamping circuit

Exercise For a 5 V peak sine wave input, sketch the output of the following circuit in steady state. Also determine the capacitor voltage in steady state.

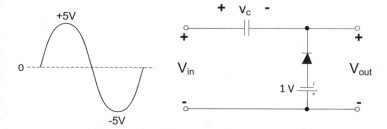

Answer The clamping voltage is −1.7 V, with the negative peak clamped to that voltage. Therefore, the output is a 5 V peak sine wave with the lower peak at −1.7 V and the upper peak at −1.7 V + 10 V = +8.3 V. The capacitor is charged to its steady-state voltage when the input is at its most negative; $v_c = -5 \text{ V} - (-1.7 \text{ V}) = -3.3 \text{ V}$.

Applications of Clamping Circuits

The clamping circuit was used extensively in the days of analog video processing, where a composite signal containing picture and timing information are embedded onto one waveform. The composite signal is transmitted over a medium where the DC value is lost, and thus the clamping circuit performs DC restoration on the signal such that the receiver can extract the different pieces from it.

A variation on the clamping concept is the voltage doubler, which generates an output DC value roughly twice that of the input. A topology known as the Dickson charge pump is shown below; the circuit appears to have two clampers in cascade, but with a clock driving the first capacitor (Fig. 16).

We will analyze the circuit assuming that V_{in} = 12 V, and CLK is a 50% duty cycle square wave varying from 0 to 5 V. Initially, both capacitors are discharged.

Fig. 16 Dickson charge pump

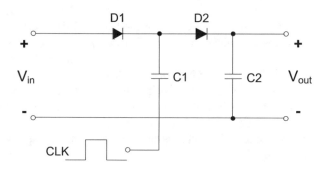

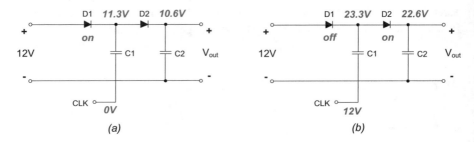

Fig. 17 Dickson charge pump operation

When CLK is at 0 V, note that D1 will turn on and charge C1 to 12–0.7 = 11.3 V. As a result, D2 also turns on to charge C2 to 11.3–0.7 = 10.6 V (Fig. 17).

Recall that the voltage across a capacitor cannot change instantaneously. When CLK jumps from 0 to 12 V, the voltage across C1 remains at 11.3 V, and hence the upper terminal of C1 jumps to 11.3 + 12 = 23.3 V. D1 is now reversed biased and D2 is forward biased, charging C2 to 23.3–0.7 = 22.6 V.

In steady state, note that when CLK = 0 V, D1 is on and D2 is off, and C1 is charged to 11.3 V. When CLK goes from 0 to 12 V, D1 is off and D2 is on. C1 provides charge to C2, bringing its voltage to 22.6 V. The sizing of C1 and C2 is not discussed here, but we can see that this circuit takes a 12 V input and provides a 22.6 V output, almost doubling the input voltage.

3 Schottky Diodes

The diodes we have worked with so far are silicon diodes, made by sandwiching p-type silicon with n-type silicon. Although this is the most common type of diodes, there are other variations that we need to be aware of. One of them is the Schottky diode, and its symbol is similar to the regular diode with some modifications. A Schottky diode is made by sandwiching metal against silicon; we will not study the device physics here, but a metal-semiconduction junction can perform rectification (Fig. 18).

One advantage of the Schottky diode is that its forward voltage is lower than a silicon diode's. Typically, we assume that 0.3 V will turn on the Schottky diode, as opposed to 0.7 V for a silicon diode. The lower forward voltage of the Schottky diode leads to better efficiency. For example, consider the circuits below, where the diode is forward biased in both cases. We can see that the power loss in the silicon diode is $I \cdot 0.7$ V, whereas for the Schottky diode it is $I \cdot 0.3$ V. The lower power loss in the Schottky diode results in better efficiency (Fig. 19).

We can also see this in the performance of the Dickson charge pump. Using Schottky diodes instead, the output voltage is now 23.4 V as opposed to 22.6 V with silicon diodes. The output is closer to doubling the input voltage compared to before (Fig. 20).

Fig. 18 Schottky diode (**a**) symbol (**b**) construction

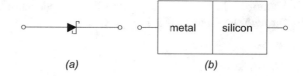

(a)   *(b)*

Fig. 19 Power loss (**a**) silicon diode (**b**) Schottky diode

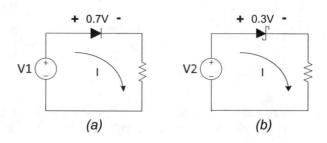

(a)   *(b)*

Fig. 20 Dickson charge pump with Schottky diodes

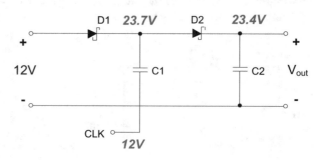

Fig. 21 Zener diode (**a**) in forward bias (**b**) in reverse bias under breakdown

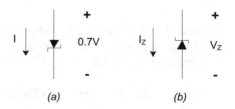

(a)   *(b)*

4   Zener Diodes

This device is a diode that has been engineered to have a low reverse breakdown voltage. In typical use, the Zener diode is operated in breakdown under reverse bias, generating a voltage that is known as the Zener voltage (Fig. 21).

Note that the Zener diode symbol looks similar to the Schottky diode symbol, but with a small difference. The operation of a Zener diode is illustrated by the following examples.

Example In the circuit below, the Zener diode has a $V_Z = 3.3$ V. What is the current through the diode? What is V_{out}?

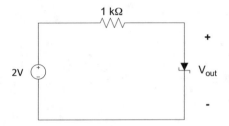

Answer Since the Zener diode is forward biased, we can treat it like a regular diode; $V_{out} = 0.7$ V, and the diode current is $(2-0.7 \text{ V})/1 \text{ k}\Omega = 1.3$ mA.

Example In the circuit below, the Zener diode has a $V_Z = 3.3$ V. What is the current through the diode? What is V_{out}?

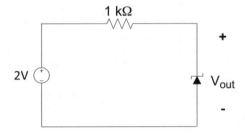

Answer The Zener diode is in reverse bias, but the applied voltage of 2 V is less than the Zener voltage. Hence, the diode is *not* in reverse breakdown, and there is no current through the diode. By KVL, $V_{out} = 2$ V.

Example In the circuit below, the Zener diode has a $V_Z = 3.3$ V. What is the current through the diode? What is V_{out}?

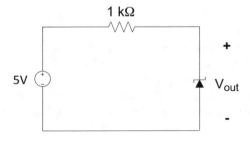

Answer The Zener diode is in reverse bias, and the applied voltage is higher than the Zener voltage. The diode will enter reverse breakdown and conduct current, with $V_{out} = V_Z = 3.3$ V. The diode current is (5–3.3 V)/1 kΩ = 1.7 mA

Example In the previous example, if we increase the DC power supply from 5 to 6 V, what is V_{out} equal to?

Answer V_{out} is still equal to 3.3 V, but the current increases to (6–3.3 V)/1 kΩ = 2.7 mA.

A Simple Voltage Regulator

We can now see how a Zener diode can be used to generate a fixed DC voltage. This can be used as a voltage reference or as another DC power supply; the Zener diode can *regulate* this output voltage over a range of load currents and input voltages. The following example will illustrate these ideas.

Example Using a 9 V battery and a 5 V Zener diode, determine that range of loads that can be used in the regulator below.

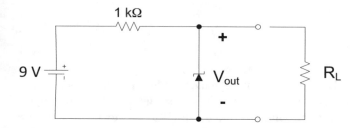

Answer Let us consider what the output voltage and diode current are in the case of no load. The Zener diode is conducting in reverse breakdown, and so $V_{out} = 5$ V. The diode current is (9 – 5 V)/1 kΩ = 4 mA.

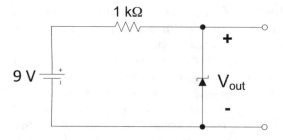

If a 100 Ω load is attached, will the output still be 5 V? We can check V_{out} with the diode removed, and see if V_{out} is big enough to cause the Zener diode to conduct in the reverse direction. In this case, with no diode, the load forms a voltage divider with the 1 kΩ resistor, giving an output of 9 V * 100 Ω / (1 kΩ + 100 Ω) = 0.818 V. This voltage is not enough to cause the Zener diode to enter reverse breakdown; hence even after the diode to reconnected to the circuit, the current is still zero and V_{out} = 0.818 V. The output voltage is no longer regulated at 5 V for a 100 Ω load.

Question What is the range of loads that this regulator can support?

Answer Let us suppose there is some current through the Zener diode, such that it is in reverse breakdown and V_{out} = 5 V. The current through the 1 kΩ resistor is then 4 mA, as found earlier for the no-load case. As long as the load current is less than 4 mA, there will be a non-zero current through the diode. The heaviest load that this regulator can accommodate is 5 V/4 mA = 1.25 kΩ. The range of loads this regulator can handle is 1.25 kΩ < R_L < ∞.

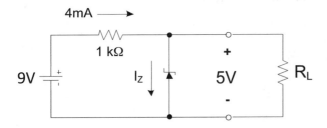

We can now see one limit of this simple regulator: the load current must be lower than the *idle* current through the 1 kΩ resistor. If we want a higher load current, we must increase the idle current, which is not an efficient way to build a voltage regulator.

5 Linear Regulators

Zener regulators are simple, but they suffer from a limited load range. If we increase the circuit complexity, we can obtain a much wider range. Luckily, we do not have to deal with that complexity; we can use an integrated circuit known as linear regulators, which are typically three terminal devices that are very easy to use. A popular example is the 78xx series, which offers a variety of regulated voltages depending on the part number. For example, the 7805 provides a 5 V regulated output voltage.

For a linear regulator to work, the input voltage must be higher than the output voltage. An important parameter of a linear regulator is the **dropout voltage** V_{DROP}. For the regulator to function properly, the input voltage must be higher than the output by at least V_{DROP} at all times.

Example A 7805 linear regulator is specified to have a dropout voltage of 2 V. What is the minimum voltage required at the input?

Answer The 7805 provides a 5 V regulated output. With a dropout voltage of 2 V, the minimum input voltage is $5 + 2 = 7$ V (Fig. 22).

A linear regulator uses negative feedback to maintain its output at the desired voltage. Shown below is a simplified schematic of a typical linear regulator. A scaled copy of the output is taken from the voltage divider R1 and R2, going to the negative input of the op-amp. The op-amp compares this to the reference on its positive input, and the op-amp drives the NPN base such that V_{cut} is maintained at the desired level, for example, 5 V in the case of the 7805. Note that the NPN transistor is operating in the active region (Fig. 23).

To see how this works, an ideal op-amp in negative feedback maintains the same voltage at its positive and negative inputs. By KVL, we see that the voltage across R2 is equal to $V_{REF} = 1.25$ V, which in turn results in a current $I = 1.25$ V/R2. Given that the input current of an ideal op-amp is zero, this current I also flows through R1. As a result, $V_{OUT} = I \times (R1 + R2)$, and therefore

$$V_{OUT} = 1.25\,\text{V} \times \left(1 + \frac{R1}{R2}\right)$$

Fig. 22 Linear regulator and the dropout voltage

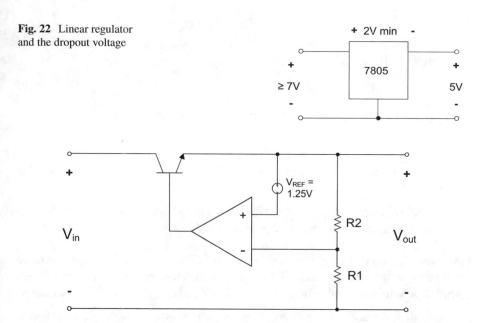

Fig. 23 Linear regulator, simplified schematic

By adjusting the ratio R1/R2, we can change the value of V_{OUT} as long as the NPN transistor is in the active region (i.e., the dropout voltage requirement is satisfied). In regulators such as the 7805, R1 and R2 are inside the IC, setting the output to 5 V. Another linear regulator, called the LM317, requires the user to use external resistors for R1 and R2 and provides the flexibility to set your own regulated output voltage.

Example We wish to use a LM317 linear regulator to create an output voltage of 3.3 V. The input voltage is sufficient to satisfy the dropout voltage requirement. If R2 is equal to 1 kΩ, what value of R1 is needed?

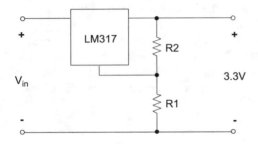

Answer We can use the equation provided earlier to solve for R1:

$$V_{OUT} = 1.25\,\text{V} \times \left(1 + \frac{R1}{1\,k\Omega}\right) = 3.3\,\text{V}$$

This gives R1 = 1.64 kΩ.

Regulated DC Power Supply

A linear regulator can be incorporated into the filtered rectifier to remove the residual ripple voltage and allow us to create a DC power supply with a clean, low noise output (Fig. 24).

Example Using a 12 V transformer, design a 5 V DC power supply with 0.5A full load. Assume that the dropout voltage of the 7805 is 2 V.

Answer The transformer secondary voltage is 12 V_{RMS}, or 17.0 V_{PK}. This sine wave is then input to a bridge rectifier (D1 to D4) and then filtered by the capacitor C. We need to determine the value of C to ensure that V_{OUT} is regulated to 5 V at all times.

The filtered rectifier output goes to the regulator input, which is shown below with and without a capacitor. $V_{PK} = 17\,\text{V} - 2 \times 0.7\,\text{V} = 15.6\,\text{V}$.

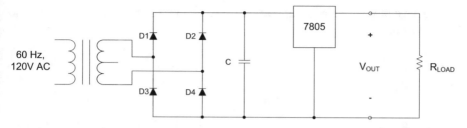

Fig. 24 Regulated DC power supply

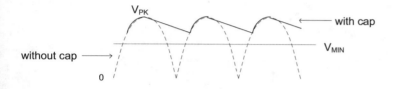

V_{MIN} is the minimum voltage that we need at the regulator input, which is equal to the output voltage plus the dropout voltage. Hence, $V_{MIN} = 5$ V $+ 2$ V $= 7$ V, so the rectifier output must always be above 7 V. In turn, the maximum ripple $\Delta v = 15.6$ V $- 7$ V $= 8.6$ V. Finally, recall that

$$\Delta v = \frac{I_{LOAD}}{Cf}$$

I_{LOAD} in this equation refers to the load current drawn from the filtered rectifier and is the same as the regulator input current. How is the regulator input current related to the regulator output current? In general, the *input and output currents are the same* for a linear regulator (i.e., $I_{in} = I_{out}$).

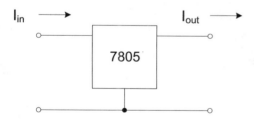

In this example, the full load current drawn from the regulator output is 0.5A. The maximum input current to the 7805 is also 0.5 A, and solving for C, we arrive at

$$C = \frac{I_{LOAD}}{\Delta v \cdot f} = \frac{0.5A}{8.6V \cdot 120Hz} = 484 \mu F$$

To have more margin in our design, we can choose a higher capacitor value to reduce ripple.

6   Problems

1. Clipping Circuits: A 5 V peak sine wave is applied to the input of the circuits below. Sketch the output waveform for each.

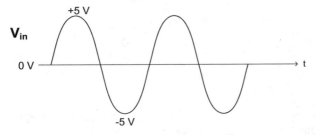

(a)

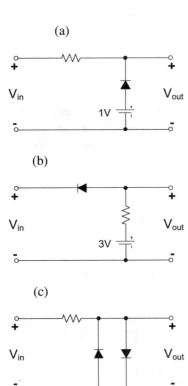

(b)

(c)

(d)

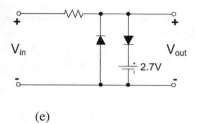

(e)

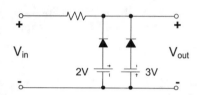

2. Clamping Circuits: A 5 V peak sine wave is applied to the input of the circuits below. Sketch the output waveform for each.

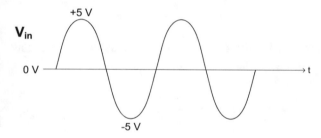

(a)

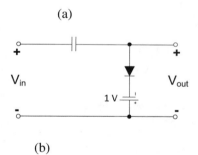

(b)

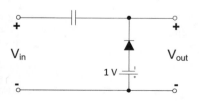

3. Suppose that a 5 V peak sine wave is applied to the input of a simple clamping circuit and allowed to reach steady state. If the input is suddenly increased to a 10 V peak sine wave, what would the output look like in steady state?

4. Suppose that a 5 V peak sine wave is applied to the input of a simple clamping circuit and allowed to reach steady state. If the input is suddenly increased to a 2 V peak sine wave, what would the output look like in steady state?

5. In practical clamping circuits, we must consider a discharge path for the capacitor such that it can respond to changes in signal amplitude. Consider a 1 kHz, 5 V peak *square* wave applied to the circuit below, with a resistor added for capacitive discharge. Sketch the output of the clamping circuit for the component values shown below.

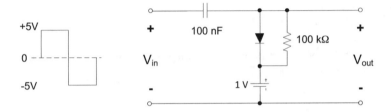

6. Dickson Charge Pump: Predict the output voltage V_{out} assuming that the Schottky diode forward voltage is 0.3 V.

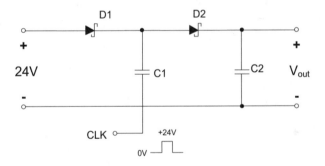

7. A Zener diode with $V_Z = 3.3$ V is used in the circuit below. Find V_{OUT} and the current thru the Zener diode for the following cases:

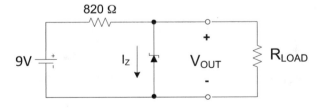

(a) $R_{\text{LOAD}} \rightarrow \infty$

(b) $R_{\text{LOAD}} = 1 \text{ k}\Omega$

(c) $R_{\text{LOAD}} = 220 \ \Omega$

8. The LM317 is a linear regulator that uses external resistors to set the output voltage, giving the user some flexibility in this regard. The voltage is set according to the following equation:

$$V_{\text{out}} = 1.25\text{V}\left(1 + \frac{R2}{R1}\right)$$

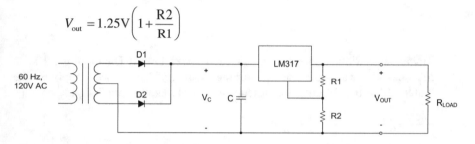

You are to design a DC power supply that provides a regulated 3.3 V output, with a full load current of 200 mA. A 12 VCT transformer is used.

(a) If R1 = 1 kΩ in the LM317 circuit, find the value of R2 that will create a 3.3 V output.

(b) If the dropout voltage for the LM317 is 1.7 V, find the minimum capacitor value that is required for this design.

9. As discussed in this chapter, a block diagram of a linear regulator is shown below.

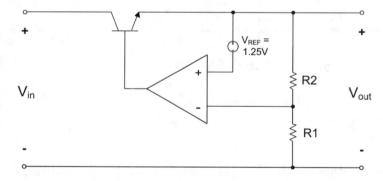

The NPN transistor in series between the input and output is sometimes called a pass transistor. Why must this NPN be operating in the active region?

10. Tinkercad®: Simulate the clipping circuit shown below. The input is a 1 kHz, 5 V peak sine wave, with the resistor set to 100 kΩ. Use a power supply for the battery. Measure the output on a scope and verify against prediction.

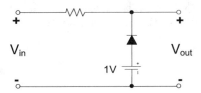

11. Tinkercad®: Simulate the clamping circuit shown below. The input is a 1 kHz, 5 V peak sine wave, and use 1 μF for the capacitor. Use a power supply for the battery. Measure the output on a scope and verify against prediction.

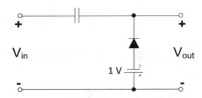

12. Tinkercad®: Simulate the Dickson Charge Pump shown below, with a clock frequency of 100 kHz. Measure the DC voltage at the output (Note: after starting the simulation, it may take several seconds for the output to rise to the final voltage).

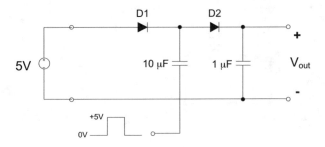

13. Tinkercad®: Build the Zener regulator below with a 3.3 V Zener diode. Measure the output voltage using a 10 kΩ load. Then decrease the load resistance until the regulator no longer works, and compare the result against theory.

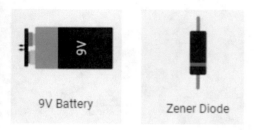

9V Battery   Zener Diode

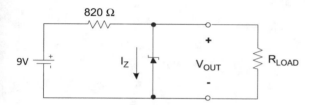

14. Tinkercad®: Using the LM7805 linear regulator, build the circuit below with a 100 Ω load. Set the input voltage to 10 V and measure the output voltage. Also note the relationship between the input current and the load current. Then decrease the input voltage until the regulation fails, which will give the dropout voltage for this regulator.

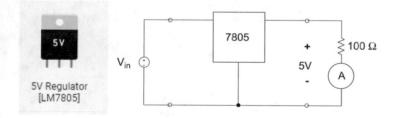

5V Regulator
[LM7805]

Diode and BJT Equations

Diodes and transistors are nonlinear devices, meaning that their I-V curves are not straight lines. We have used simple models for the diode and BJT to facilitate hand analysis. In this chapter, we will take a look at the actual equations governing the diode and BJT and the use of graphical methods to find solutions for nonlinear devices.

1 Diodes and the Shockley Equation

The model that we have used for the diode is a simple one, in which we have a fixed 0.7 V forward bias if the diode is on. This model is sufficient for most analysis, but there will be cases when a more detailed description is needed. The I-V curve of a diode is described more accurately with Shockley's equation:

$$I_D = I_S \cdot \left(e^{V_D/V_T} - 1 \right)$$

- I_S = saturation current
- V_T = thermal voltage = $kT/q \approx 26$ mV at $T = 27\ °C$
- k = Boltzmann's constant = 1.38×10^{-23} J/K
- T = temperature in Kelvin
- Q = electron charge = 1.6×10^{-19} C

Note that the diode current I_D increases exponentially with increasing forward bias voltage V_D (Fig. 1).

Note that in forward bias, where we had used $V_D = 0.7$ V in the simple model, the ratio V_D/V_T is typically large. Therefore, $\exp(V_D/V_T)$ is much greater than 1, and Shockley's equation can be simplified to

© Springer Nature Switzerland AG 2022
C. Siu, *Electronic Devices, Circuits, and Applications*,
https://doi.org/10.1007/978-3-030-80538-8_11

Fig. 1 Diode I-V curve: simple model vs Shockley equation

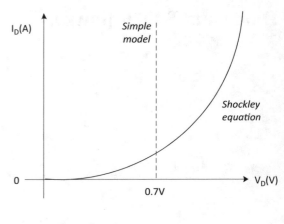

$$I_{\mathrm{D}} = I_{\mathrm{S}} \cdot \left(e^{V_{\mathrm{D}}/V_{\mathrm{T}}} \right)$$

On the other hand, in reverse bias the ratio $V_{\mathrm{D}}/V_{\mathrm{T}}$ is a large negative number, and $\exp(V_{\mathrm{D}}/V_{\mathrm{T}})$ is much less than 1. Shockley's equation can be approximated as shown below:

$$I_{\mathrm{D}} = -I_{\mathrm{S}}$$

The saturation current I_{S} is a small value, but it is nonzero. Hence, a real diode in reverse bias conducts a small current, but in most applications this can be ignored.

Example The diode below has $I_{\mathrm{s}} = 1$ nA at $T = 27$ °C. Find V and I in the circuit below:

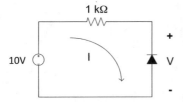

Answer The diode is reverse biased; hence the current $I = -I_{\mathrm{D}} = I_{\mathrm{S}} = 1$ nA. The voltage drop across the 1 kΩ resistor is only 1 µV, which can be neglected in most cases. The voltage V is thus 10 V.

Example The diode below has $I_{\mathrm{s}} = 1$ nA at $T = 27$ °C. Find V and I in the circuit below:

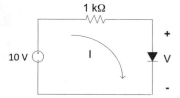

Answer In this case, the diode is forward biased, and with the simple model, the analysis is straightforward: $V = 0.7$ V and $I = 9.3$ mA.

If we wish to use Shockley's equation, however, we need to solve a system of two equations:

$$I = I_S \cdot \left(e^{V/V_T} \right)$$

$$I = \frac{10 - V}{1\ k\Omega}$$

The second equation is obtained by applying Ohm's law to the 1 kΩ resistor. To solve for V, we have a nonlinear equation which is difficult to solve by hand.

$$\frac{10 - V}{1k\Omega} = (1nA) \cdot \left(e^{V/V_T} \right)$$

Although computers can be used to solve this using numerical methods, we can also estimate the solution by graphing the two equations.

Graphical Method of Finding DC Operating Points

To illustrate how we can solve nonlinear equations graphically, we begin with an example of solving two linear equations.

Example Given the equations below, solve for X and Y using graphic methods

$$X + Y = 3$$

$$2X - 3Y = 10$$

Answer Both equations can be plotted on an X-Y graph. The intersection of the two lines is the solution to the equations, which can be read off as approximately $X = +4$, $Y = -1$ (the exact solution is $X = +3.8$, $Y = -0.8$)

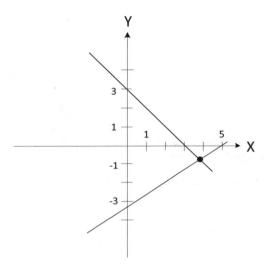

We will extend this method to functions that are not linear.

Example Shown below is the measured I-V curve for a 1N4005 diode.

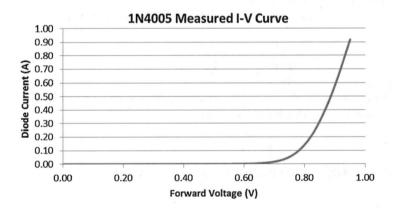

For the circuit below, find the current I_D and voltage V_D.

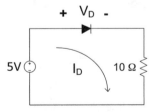

Using the simple model, $V_D = 0.7$ V and $I_D = 0.43$ A.

Next, we use the measured characteristics of the diode to find the solution. This is done by finding I_D through the 10 Ω resistor using Ohm's law.

$$I_D = \frac{5 - V_D}{10\Omega} = -0.1 \cdot V_D + 0.5$$

Note that this is a linear equation on an I_D versus V_D graph. This equation creates what is known as a *load line* on the graph, for which we can regard the 10 Ω resistor as a load on the circuit. Plotting the load line on the same graph as the 1N4005 I-V curve, we can find the circuit solution by noting the intersection between the two.

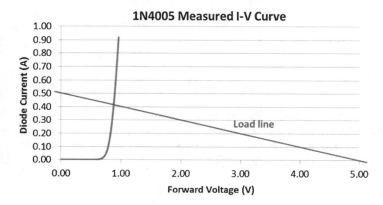

Zooming into the graph, we can get a better estimate of the solution.

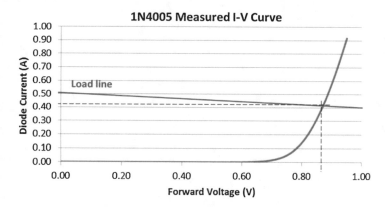

The solution is approximately $I_D = 0.42$A, $V_D = 0.87$ V. Note that the current is quite close to the result obtained using the simple model, illustrating the utility of the simple model in obtaining solutions quickly.

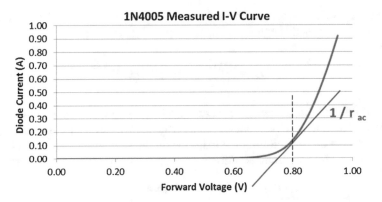

Fig. 2 1N4005 Diode I-V Curve with AC resistance at $V_D = 0.8$ V

AC Resistance of a Diode

We can also use the measured diode characteristics to estimate the diode's AC resistance, by calculating the slope at different points on the curve (Fig. 2).

Shown here is a subset of the 1N4005 measurements. We can now estimate the AC resistance for various DC operating points.

Vd (V)	Id (A)
0.6	0.001549
0.61	0.002003
0.62	0.002515
0.63	0.003026
0.64	0.003992
0.65	0.005186
0.66	0.006493
0.67	0.008141
0.68	0.010187
0.69	0.013029
0.7	0.016609
0.71	0.020588
0.72	0.026214
0.73	0.032863
0.74	0.041104
0.75	0.051334
0.76	0.06378
0.77	0.0785
0.78	0.096232
0.79	0.116975
0.8	0.141015

Vd (V)	Id (A)
0.81	0.168352
0.82	0.199496
0.83	0.234448
0.84	0.273151
0.85	0.315548

Example Estimate the AC resistance of the 1N4005 for $V_D = 0.7$ V.

Answer To find the slope at $V_D = 0.7$ V, we can use the two nearest data points at $V_D = 0.69$ V and 0.71 V.

$$r_{ac} = \frac{0.71 \text{ V} - 0.69 \text{ V}}{20.6 \text{ mA} - 13.0 \text{ mA}} = 2.63 \ \Omega$$

Example Estimate the AC resistance of the 1N4005 for $V_D = 0.8$ V.

Answer To find the slope at $V_D = 0.8$ V, we can use the two nearest data points at $V_D = 0.79$ V and 0.81 V.

$$r_{ac} = \frac{0.81 \text{ V} - 0.79 \text{ V}}{0.168 \text{ A} - 0.117 \text{ A}} = 0.39 \ \Omega$$

If measured data is not available, we can still estimate a diode's AC resistance by using Shockley's equation. In forward bias, recall that

$$I_D = I_S \cdot \left(e^{V_D/V_T} \right)$$

To find the slope at different points on the I-V curve, we calculate the derivative of Shockley's equation:

$$\frac{dI_D}{dV_D} = I_S \cdot \left(e^{V_D/V_T} \right) \cdot \frac{1}{V_T} = \frac{I_D}{V_T}$$

Since the AC resistance is the inverse of the slope, a diode's AC resistance is given by the following expression:

$$r_{ac} = \left(\frac{dI_D}{dV_D} \right)^{-1} = \frac{V_T}{I_D}$$

Example The diode below has $I_s = 1$ nA at $T = 27$ °C. Find the diode's AC resistance.

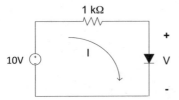

Answer We can use the simple model to estimate the DC operating point, which is $V = 0.7$ V and $I = 9.3$ mA. Next, at $T = 27$ °C or 300 K, V_T is about 26 mV; hence $r_{ac} = 26$ mV/9.3 mA = 2.8 Ω.

Example Shown below is a diode circuit containing a DC source and an AC source. Sketch the voltages across the diode and resistor as a function of time. Estimate the diode AC resistance from theory, given that $I_S = 1$ nA.

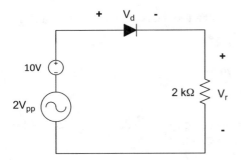

Answer We can use superposition to analyze this circuit. Considering the DC source only, we can estimate the DC operating point.

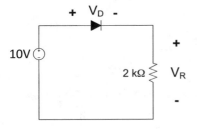

Since the diode is forward biased, $V_D = 0.7$ V, $V_R = 9.3$ V, and $I_D = 9.3$ V/2kΩ = 4.65 mA. The diode AC resistance is 26 mV/4.65 mA = 5.6 Ω.

Next, we consider the AC source only and replace the diode by its AC resistance. The input is a 1 V peak sine wave, and the voltage divides between the diode and the resistor. vd and vr are both sinusoidal, with values of 2.79 mV peak and 0.997 V peak, respectively.

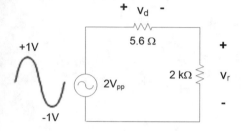

Finally, we add the DC and AC results together. For example, the resistor voltage waveform V_r has a 9.3 V average value, with a 0.997 V peak sine wave varying about this 9.3 V average.

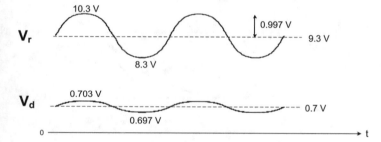

2   The BJT Equation

We have assumed that when the BJT is on, $V_{BE} = 0.7$ V. Just like the diode, this is an approximation that is usually sufficient for hand analysis. In reality, V_{BE} does vary as the collector current changes, and the relationship between I_C and V_{BE} is governed by (Fig. 3)

$$I_C = I_S \cdot \left(e^{\left(V_{BE} / V_T \right)} \right)$$

Hence for a BJT in the active region, the base current I_B is related to the collector current I_C via β. At the same time, V_{BE} varies as the natural log of I_C; even if I_C changes by a factor of 10, V_{BE} only changes by 60 mV at room temperature.

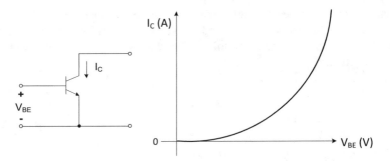

Fig. 3 BJT I_C vs V_{BE} characteristic in the active region

Fig. 4 Find the transconductance at operating point I_C

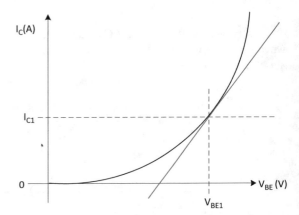

The Hybrid-Pi Model

Similar to the NMOS, we can develop a model for the NPN in the active region, based on the small signal assumption. In contrast to the NMOS, however, the AC resistance of the base is not infinite, so we need to account for this in the model.

First, the definition of transconductance g_m is similar to the NMOS: we take the device equation for I_C vs V_{BE} and differentiate it to find the slope at a specific DC operating point (Fig. 4):

$$g_{\mathrm{m}} = \frac{dI_{\mathrm{C}}}{dV_{\mathrm{BE}}} = I_{\mathrm{s}} \cdot \left(e^{V_{BE}/V_{\mathrm{T}}} \right) \cdot \frac{1}{V_{\mathrm{T}}} = \frac{I_{\mathrm{C}}}{V_{\mathrm{T}}}$$

As we vary V_{BE}, the collector current I_C and base current I_B will also change, related by $I_C = \beta \cdot I_B$ in the active region. This leads to a finite AC resistance between the base and emitter of a BJT (Fig. 5).

Fig. 5 Finding the base
AC resistance

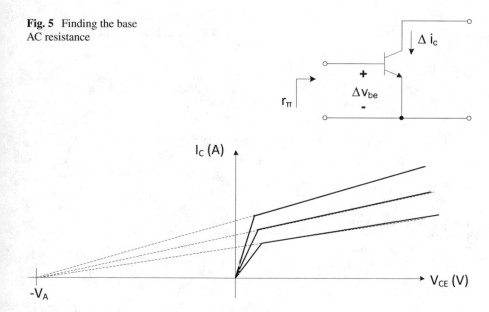

Fig. 6 Collector AC resistance and the early voltage

$$I_B = \frac{I_C}{\beta} = \frac{I_S}{\beta} \cdot \left(e^{V_{BE}/V_T} \right)$$

$$\frac{dI_B}{dV_{BE}} = \frac{I_S}{\beta} \cdot \left(e^{V_{BE}/V_T} \right) \cdot \frac{1}{V_T} = \frac{I_B}{V_T}$$

Thus, we can find the base AC resistance r_π by taking the inverse of the derivative:

$$r_\pi = \left(\frac{dI_B}{dV_{BE}} \right)^{-1} = \frac{V_T}{I_B}$$

Finally, the AC resistance seen between the collector and emitter is also finite. On an I_C vs V_{CE} graph, this can be calculated using the slope in the active region. However, we can make this more general by extrapolating the slopes of multiple I-V curves back to the horizontal axis (Fig. 6):

All the extrapolated slopes meet at $-V_A$, where V_A is known as the *early voltage*, named after American engineer Jim Early. Early was the first person to characterize this effect for BJTs in a 1952 paper. The collector AC resistance can be calculated using the following equation:

$$r_\circ = \frac{V_A}{I_C}$$

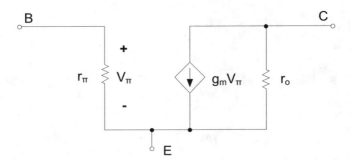

Fig. 7 NPN hybrid-Pi model

Fig. 8 Finding the emitter
AC resistance

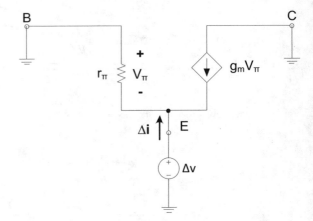

All the pieces are now in place to construct a small signal model for a NPN transistor in the active region. This is commonly known as the *hybrid-pi model*. Note that this model looks very similar to the NMOS small signal model, with the exception of the base resistance r_π. We will use this model to analyze three single transistor amplifiers: the common emitter, common base, and common collector amplifiers (Fig. 7).

Using the hybrid-pi model, we can find the emitter AC resistance. To simplify the initial analysis, we will assume that $r_o \to \infty$. We apply a test voltage Δv to the emitter and find Δi; the ratio $\Delta v/\Delta i$ will give the AC resistance seen at the emitter (Fig. 8).

From the equivalent circuit, we see that $\Delta v = -v_\pi$ and $\Delta i = -(v_\pi/r_\pi + g_m v_\pi)$. Combining these two equations we get the following:

$$r_e = \frac{\Delta v}{\Delta i} = \left(\frac{1}{r_\pi} + g_m \right)^{-1}$$

This equation is commonly written in other ways. Noting that $g_m r_\pi = (I_C/V_T)$ $(V_T/I_B) = \beta$, or $1/r_\pi = g_m/\beta$:

$$r_e = \left(\frac{g_m}{\beta} + g_m \right)^{-1} = \left(\frac{\beta+1}{\beta} g_m \right)^{-1}$$

Defining the common base current gain α as,

$$\alpha = \frac{I_C}{I_E} = \frac{\beta}{\beta+1}$$

The emitter AC resistance can be written as

$$r_e = \left(\frac{g_m}{\alpha} \right)^{-1} = \frac{\alpha}{g_m} \cong \frac{1}{g_m}$$

The approximation $\alpha \approx 1$ is valid if $\beta \gg 1$, which is usually the case. In summary, the AC resistances of a NPN in the active region are shown below (Fig. 9):

$$r_\pi = \frac{V_T}{I_B} = \frac{V_T \cdot \beta}{I_C} = \frac{\beta}{g_m}$$

$$r_e = \frac{V_T}{I_E} = \frac{\beta}{\beta+1} \cdot \frac{V_T}{I_C} = \frac{\alpha}{g_m}$$

$$r_o = \frac{V_A}{I_C}$$

Beta Reflection Rule

We saw that the AC base resistance is r_π if the emitter is connected to ground. If the emitter is not grounded but connected to a resistor instead, does that change the base resistance? We will use the hybrid-pi model, apply a test voltage Δv at the base, and find Δi; the ratio $\Delta v/\Delta i$ will give us the AC resistance (Fig. 10).

The voltage across the emitter resistor R is given by

$$v_e = \left(\Delta i + g_m v_\pi \right) \cdot R$$

By KVL, we find that

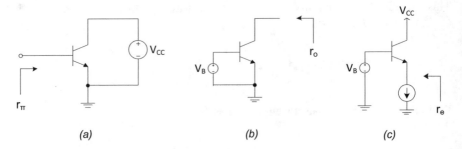

Fig. 9 AC resistances of a NPN transistor in the active region

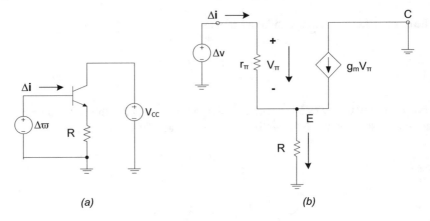

Fig. 10 Finding the base AC resistance with an emitter resistor R

$$\Delta v = v_\pi + v_e = v_\pi + \left(\Delta i + g_m v_\pi\right) \cdot R$$

Also, since $v_\pi = \Delta i \cdot r_\pi$, we can substitute this into the equation above to find the AC resistance:

$$\Delta v = v_\pi \left(1 + g_m R\right) + \Delta i \cdot R = \Delta i \cdot r_\pi \left(1 + g_m R\right) + \Delta i \cdot R$$

$$r_{in} = \frac{\Delta v}{\Delta i} = r_\pi \left(1 + g_m R\right) + R = r_\pi + \left(\beta + 1\right) \cdot R$$

Note that the base resistance is not r_π, but is increased by an additional term $(\beta + 1) \cdot R$. The resistance in the emitter appears to be $\beta + 1$ larger; this is known as the beta reflection rule, where an emitter resistance will appear to be much larger when viewed from the base.

3 BJT Amplifiers

Using the hybrid-pi model, we can analyze the various single BJT amplifiers:

- Common emitter amplifier
- Common base amplifier
- Common collector amplifier or emitter follower

Common Emitter Amplifier

The common emitter amplifier is analogous to the MOSFET common source amplifier, where the emitter terminal is bypassed to AC ground. The input AC signal is fed
into the base, and the output is taken from the collector (Fig. 11).

First, we can represent this amplifier with an AC equivalent circuit, similar to the
techniques we used for the common source amplifier: replace the transistor with its
small signal model, and assume that all capacitors are AC short circuits (Fig. 12).

Defining the input resistance r_{in} and output resistance r_{out} as

$$r_{\text{in}} = R_{\text{B1}} \parallel R_{\text{B2}} \parallel r_{\pi}$$

$$r_{\text{out}} = R_{\text{C}} \parallel r_{\text{o}}$$

Using the AC equivalent circuit, we can see that from v_{in} to v_{π}, there is a voltage
division:

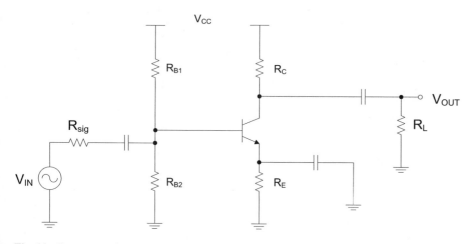

Fig. 11 Common emitter amplifier

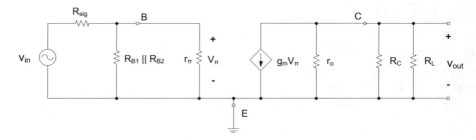

Fig. 12 AC equivalent circuit, CE amplifier

$$v_\pi = \frac{r_{in}}{R_{sig} + r_{in}} v_{in}$$

At the output, v_{out} is due to the current from the dependent source flowing through the parallel resistors r_o, R_C, and R_L:

$$v_{out} = -g_m v_\pi \left(r_{out} \| R_L \right)$$

Combining the two equations, we can find the voltage gain v_{out}/v_{in}:

$$A_v = \frac{v_{out}}{v_{in}} = -\frac{r_{in}}{R_{sig} + r_{in}} g_m \left(r_{out} \| R_L \right)$$

Note that the input resistance and output resistance of the CE amplifier are correctly defined by r_{in} and r_{out} above. As seen the in figure below, we are interested in the amplifier input resistance from the signal generator's point of view (v_{in} and R_{SIG}). At the output, the load is not included in r_{out} since the load is not intrinsic part of the amplifier; the amplifier drives the external load R_L, with its attendant effect on the amplifier gain (Fig. 13).

Example Find the gain, input resistance, and output resistance of the common emitter amplifier below.

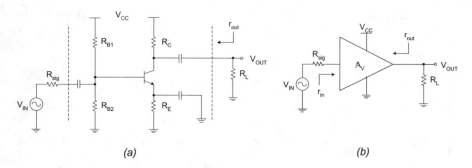

Fig. 13 Common emitter amplifier, input and output resistances

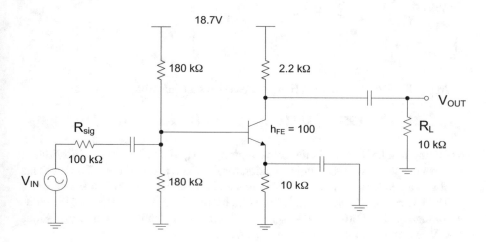

Step 1: Find the DC operating point

Note that there is a voltage divider composed of 180 kΩ resistors connected to the base. Due to the finite base current, we cannot simply apply the voltage divider rule. One analysis approach is to find the Thevenin equivalent of the voltage divider first:

$$V_{TH} = \frac{180 \ k\Omega}{180 \ k\Omega + 180 \ k\Omega}(18.7 \ V) = 9.35 \ V$$

$$R_{TH} = 180 \ k\Omega \, || \, 180k\Omega = 90k\Omega$$

V_{TH} and R_{TH} are the Thevenin voltage and resistance of the divider. We can now redraw the schematic as shown below.

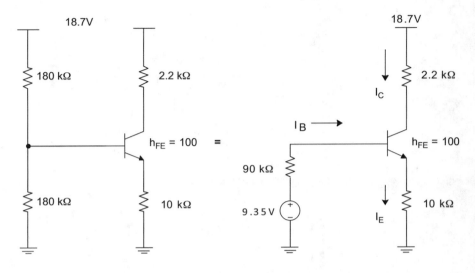

Apply KVL to the loop involving the base-emitter junction,

$$-9.35\text{V} + \left(90 \ k\Omega\right)\cdot I_B + 0.7 \ \text{V} + \left(10 \ k\Omega\right)\cdot I_E = 0$$

Assuming that the NPN is in the active region, $I_E = (\beta + 1)\cdot I_B = 101\cdot I_B$. Substituting this into the above yields a single equation for I_B, which is found to be 7.86 µA. This results in a collector current $I_C = 0.786$ mA and an emitter current $I_E = 0.794$ mA.

The DC collector voltage is 18.7 V – (2.2 kΩ)(0.786 mA) = 17.0 V, and the emitter voltage is (10 kΩ)(0.794 mA) = 7.94 V. The collector-emitter voltage V_{CE} is thus 9.06 V, and the NPN is in the active region as assumed.

Step 2: Calculate the transistor small signal parameters. The Early voltage V_A for this NPN is 50 V.

$$g_m = \frac{0.786 \ \text{mA}}{26 \ \text{mV}} = 30.2 \ \text{mS}$$

$$r_\pi = \frac{26 \ \text{mV}}{7.86 \ \mu A} = 3.31 \ k\Omega$$

$$r_o = \frac{50 \ \text{V}}{0.786 \ \text{mA}} = 63.6 \ k\Omega$$

Step 3: Calculate the amplifier's gain, input resistance, and output resistance:

$$r_{in} = 180 \ k\Omega \,||\, 180 \ k\Omega \,||\, 3.31 \ k\Omega = 3.19 \ k\Omega$$

$$r_{out} = 2.2 \ k\Omega \,||\, 63.6 \ k\Omega = 2.13 \ k\Omega$$

$$A_v = -\frac{3.19 \ k\Omega}{100 \ k\Omega + 3.19 \ k\Omega}(30.2 \ mS)(2.13 \ k\Omega \parallel 10 \ k\Omega) = -1.64\frac{V}{V}$$

The overall amplifier gain is low due to the signal attenuation at the input, between the signal generator R_{sig} and the amplifier r_{in}.

Common Base Amplifier

The common base amplifier is analogous to the MOSFET common gate amplifier, where the base terminal is bypassed to AC ground. The input AC signal is fed into the emitter, and the output is taken from the collector (Fig. 14).

The AC equivalent circuit of the common base amplifier is as shown, with $r_o \to \infty$ to simplify the analysis. Note that the current through R_{sig} is equal to $g_m v_\pi + v_\pi/r_\pi$. Performing KVL around the input (Fig. 15),

$$v_\pi + \left(g_m v_\pi + \frac{v_\pi}{r_\pi}\right) \cdot R_{sig} + V_{in} = 0$$

$$v_\pi = \frac{-V_{in}}{1 + \left(g_m + \frac{1}{r_\pi}\right) \cdot R_{sig}}$$

Looking at the collector, v_{out} is equal to $-g_m \cdot v_\pi \cdot (R_c \parallel R_L)$. Substituting the above equation for v_π into this, we obtain an expression for the voltage gain.

$$A_v = \frac{V_{out}}{V_{in}} = \frac{1}{1 + \left(g_m + \frac{1}{r_\pi}\right) \cdot R_{sig}} g_m \cdot (R_C \parallel R_L)$$

This equation can be simplified with $g_m r_\pi = (I_C/V_T) \cdot (V_T/I_B) = \beta$:

$$A_v = \frac{r_\pi}{r_\pi + (g_m r_\pi + 1) \cdot R_{sig}} g_m \cdot (R_C \parallel R_L) = \frac{r_\pi}{r_\pi + (\beta + 1) \cdot R_{sig}} g_m \cdot (R_C \parallel R_L)$$

Since the amplifier input is the emitter, we know from earlier results that the AC resistance is $r_e = \alpha/g_m$. It can also be shown that $r_e = r_\pi/(\beta + 1)$, hence since $r_{in} = r_e$, the gain equation can be written as

$$A_v = \frac{r_{in}}{r_{in} + R_{sig}} g_m \cdot (R_C \parallel R_L)$$

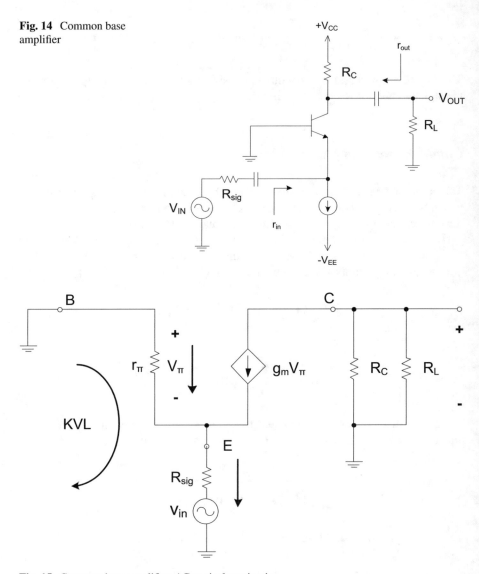

Fig. 14 Common base amplifier

Fig. 15 Common base amplifier: AC equivalent circuit

The input and output resistance of the common base amplifier is

$$r_{in} = r_e = \frac{\alpha}{g_m}$$

$$r_{out} = R_C$$

Common Collector Amplifier or Emitter Follower

The emitter follower is analogous to the MOSFET source follower, where the NPN's collector is at AC ground. The input AC signal is fed into the base, and the output is taken from the emitter (Fig. 16).

The analysis will proceed in a different route for simplicity. First, using the beta reflection rule, note that the base AC resistance is given by

$$r_b = r_\pi + (\beta + 1) \cdot R_L$$

The base AC resistance is in parallel with R_B, and so the amplifier input resistance r_{in} is

$$r_{in} = R_B \| \left(r_\pi + (\beta + 1) \cdot R_L \right)$$

Note that the signal generator v_{in} and R_{sig} forms a voltage divider with r_{in}. As a result, the AC voltage at the base is given by

$$v_b = \frac{r_{in}}{R_{sig} + r_{in}} v_{in}$$

The AC equivalent circuit for the emitter follower is shown below, assuming that $r_o \to \infty$. Note that the expression from v_{in} to v_b has already been found above. From the equivalent circuit, we obtain the following two equations (Fig. 17):

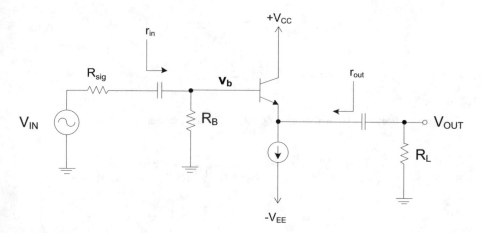

Fig. 16 Emitter follower

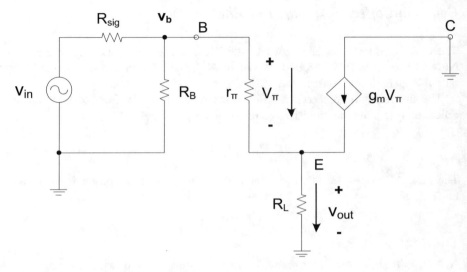

Fig. 17 Emitter follower: AC equivalent circuit

$$v_{out} = \left(\frac{v_\pi}{r_\pi} + g_m v_\pi \right) \cdot R_L = v_\pi \frac{g_m}{\alpha} R_L$$

$$v_b = v_\pi + v_{out}$$

Combining the three equations above,

$$\frac{r_{in}}{R_{sig} + r_{in}} v_{in} = v_{out} \frac{\alpha}{g_m} \frac{1}{R_L} + v_{out}$$

$$\frac{r_{in}}{R_{sig} + r_{in}} v_{in} = v_{out} \left(1 + \frac{r_e}{R_L} \right) = v_{out} \left(\frac{R_L + r_e}{R_L} \right)$$

This finally provides us with a gain equation for the emitter follower:

$$A_v = \frac{v_{out}}{v_{in}} = \left(\frac{r_{in}}{R_{sig} + r_{in}} \right) \left(\frac{R_L}{R_L + r_e} \right)$$

$$r_{in} = R_B \parallel \left(r_\pi + (\beta + 1) \cdot R_L \right)$$

$$r_{out} = r_e$$

4   Problems

1. Use the load line approach to find the DC operating point of the diode below.

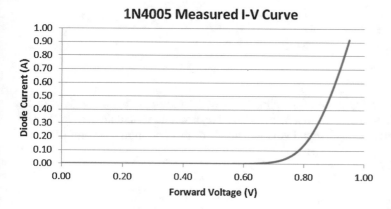

2. The table below contains the measured characteristics of a diode. Calculate the AC resistance of the diode at $V_D = 0.79$ V.

Vd (V)	Id (A)	Vd (V)	Id (A)	Vd (V)	Id (A)
0.6	0.001549	0.7	0.016609	0.8	0.141015
0.61	0.002003	0.71	0.020588	0.81	0.168352
0.62	0.002515	0.72	0.026214	0.82	0.199496
0.63	0.003026	0.73	0.032863	0.83	0.234448
0.64	0.003992	0.74	0.041104	0.84	0.273151
0.65	0.005186	0.75	0.051334	0.85	0.315548
0.66	0.006493	0.76	0.06378	0.86	0.361468
0.67	0.008141	0.77	0.0785	0.87	0.41114
0.68	0.010187	0.78	0.096232	0.88	0.464108
0.69	0.013029	0.79	0.116975	0.89	0.520428

3. A silicon diode is biased with a forward current of 0.1 mA. Calculate the AC resistance of the diode for $T = 300$ K.

4. Find the DC operating point of the NPN transistor below. Next, calculate the input resistance r_{in} and the small signal voltage gain A_V of this common emitter amplifier.

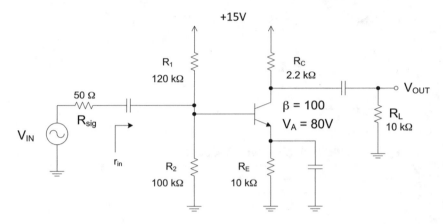

5. The DC operating point of the NPN transistor is the same as the previous question, but we will assume that $V_A \to \infty$ for this problem. Calculate the input resistance r_{in} and the small signal voltage gain A_V of this common base amplifier.

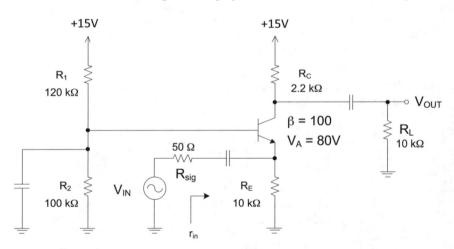

6. Find the DC operating point of the NPN transistor below. Calculate the input resistance r_{in} and the small signal voltage gain A_V of the emitter follower.

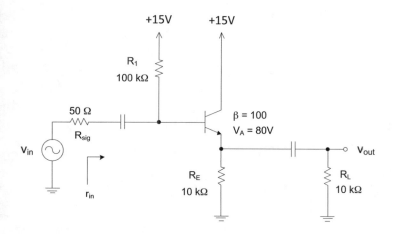

7. Tinkercad®: Construct the circuit below. Measure the diode voltage as the potentiometer is adjusted to vary the current.

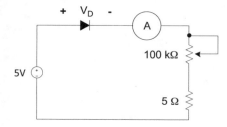

8. Tinkercad®: Construct the common emitter amplifier below. Simulate and find the NPN's DC operating point first and then use the result to calculate the small signal parameters. Predict the gain of the amplifier, comparing it against simulation (Note: after you start the simulation, it may take 10 seconds for the circuit to reach steady state due to the charging of capacitors).

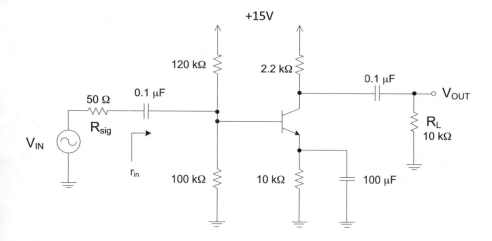

Practical Tips in Electronics

When constructing electronic circuits, we need to deal with real component behavior. Even something as simple as a wire has parasitic resistance, capacitance, and inductance. Connecting a circuit to a power supply with long wires means that power and ground has a finite impedance (inductance). If the circuit suddenly draws current, then the parasitic inductance will induce noise on the power and ground voltages. Common practices are thus observed when assembling circuits to minimize these issues.

1 Component Packaging

Circuit technology has improved in giant leaps over the past few decades, and along with that is the evolution of electronic packaging. Semiconductor, be it a single transistor or a large integrated circuit, is enclosed in a protective package with pins such that it can be used on a circuit board. In general, packaging has been driven by several requirements, such as lower cost and smaller sizes.

Passive Component Packages

We will begin by broadly dividing packages into two types: *thru-hole* and *surface-mount*. Thru-hole package is the older technology, in which resistors, capacitors, and inductors have long leads attached to their bodies. Shown below are the typical dimensions of a ¼ Watt thru-hole resistor (Fig. 1).

Over time, the demand for more functionality in a smaller volume has driven the use of surface-mount devices (SMDs) or surface-mount technology (SMT) components. Early on, it was recognized that SMT component sizes need to be standardized to enable the automated assembly of circuit boards. The Joint Electron Device

© Springer Nature Switzerland AG 2022
C. Siu, *Electronic Devices, Circuits, and Applications*,
https://doi.org/10.1007/978-3-030-80538-8_12

Fig. 1 Physical dimensions of a 0.25 watt Thru-Hole resistor

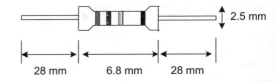

28 mm   6.8 mm   28 mm

Fig. 2 Physical dimensions of passive SMT component (**a**) top view (**b**) oblique view

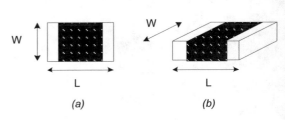

(a)   (b)

Engineering Council (JEDEC) manages and develops these standards, and for passive components such as resistors, the standard uses codes such as 0603 and 0402 for rectangular components of specific dimensions (Fig. 2, Table 1).

Comparing the size of a thru-hole resistor to that of a SMT resistor, we can see that the latter is much smaller.

Table 1 Dimensions of surface mount passive components

SMD package type	Dimensions (inches)	Dimensions (mm)
0805	0.08 × 0.05	2.0 × 1.3
0603	0.06 × 0.03	1.5 × 0.8
0402	0.04 × 0.02	1.0 × 0.5

Transistor Packages

Since transistors have three terminals, their packages must support this. For small signal transistors that do not dissipate much power, a popular thru-hole package is the TO-92 (transistor outline package) made of plastic or epoxy. For SMT, the SOT-23 (small outline transistor) is a common package for low power transistors (Fig. 3).

For power transistors, larger packages with heat sinks are used to allow higher power dissipation.

Integrated Circuit Packages

In the early days of integrated circuits, ICs were simple and did not have many inputs and outputs. Hence with the limited pin count, a common thru-hole package was the DIP (dual in-line plastic) package. This is still available today for circuits

Fig. 3 Transistor packages
(**a**) TO-92 (**b**) SOT-23

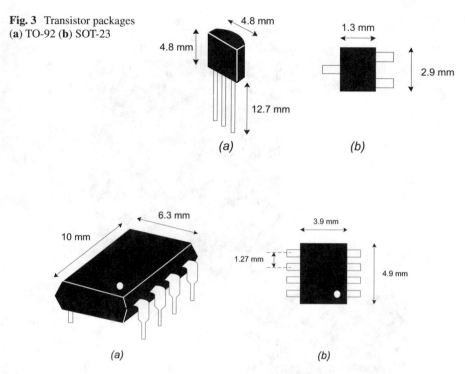

(a)   (b)

Fig. 4 IC 8-pin packages (**a**) DIP (**b**) SOIC

like operational amplifiers such as the venerable LM741 using an 8-pin DIP package.

As the push for miniaturization continued, the 8-pin packages were shrunk to a small surface mount package called SOIC (small outline integrated circuit). The dimensions of the DIP and SOIC packages are shown below for comparison (Fig. 4).

A variation on the SOIC package is the SOP (small outline package), which has a smaller pin-to-pin pitch than the SOIC. Even within the SOP family, there are different types, with two of them listed below:

- SSOP (shrink small outline package): A pin pitch of 0.65 mm
- TSSOP (thin shrink small outline package): A pin pitch of 0.5 mm and a smaller thickness compared to the SOIC

The SOIC and SOP packages are similar in that they are rectangular, with pins on two sides of the rectangle. For integrated circuits with a small number of inputs and outputs, this is a reasonable arrangement. As ICs became more complicated with hundreds of I/Os, however, this configuration would require packages that are excessive in length.

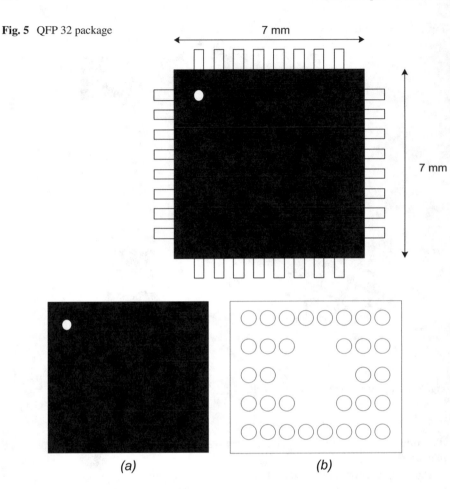

Fig. 5 QFP 32 package

Fig. 6 BGA package (**a**) top view (**b**) bottom view

The next progression is to make the package square, then put pins on all four sides. This is known as the QFP (quad flat pack) package, with pin pitches from 0.4 to 0.8 mm. A 32 pin QFP package is shown below (Fig. 5).

We can see that this scheme also has a limit. As the number of pins increase, the square must become bigger and bigger. To allow large number of I/Os in a small space, package technology moved onto the BGA (ball grid array). In this package, the connections to the ICs are made using metallic balls on the underside of the package. This scheme allows us to fully utilize the area of the square for I/Os (Fig. 6).

There are other advances in package technology that is not discussed here, such as chip scale packaging (CSP), which restricts the package size to be not much more than the size of the bare silicon IC itself.

2 Capacitor Types

Capacitors are widely used, and there are various types of made of different materials and technologies. Several of the technologies in use today are as shown.

- Electrolytic
- Polymer Film
- Ceramic

Electrolytic capacitors have the advantage of larger capacitance values in a reasonable size and low cost. Electrolytic capacitors are polarized, where the cathode is marked with minus signs along the package as shown below. If electrolytic capacitors are used with the wrong voltage polarity (i.e., cathode more positive than anode), they may fail and explode.

In aluminum electrolytic capacitors, aluminum is used as the anode, and an electrolyte is used as the cathode. The electrolyte may be solid or nonsolid. When the aluminum is in contact with the electrolyte with a positive voltage, a thin layer of oxide is formed that serves as the dielectric. If the incorrect voltage polarity is put on the aluminum, the oxide can disappear and allow conduction through the capacitor (Fig. 7).

On the other hand, ceramic capacitors use ceramic as the dielectric material. To increase the capacitance per unit volume, alternating layers of metal and ceramic can be used. Multilayer ceramic capacitors, or MLCCs, are the most common capacitors in electronic equipment. Note that due to the symmetric construction, ceramic capacitors are *non-polarized*.

One thing to note is that all capacitors have an equivalent series resistance (ESR) and a self-resonance frequency (SRF). The electrodes and dielectric have losses, and this can be modeled with a resistor in series with the capacitor. Furthermore, the capacitor leads have parasitic inductance L_{par}, and this creates a resonant circuit with the capacitance. Above the SRF, the capacitor will no longer behave like one, with the parasitic inductance beginning to dominate (Fig. 8).

Note that the larger the capacitance value, the lower the SRF. We must consider this when choosing the capacitance, such as in the use of bypass capacitors.

Fig. 7 Thru-Hole
capacitors (**a**) electrolytic
(**b**) ceramic

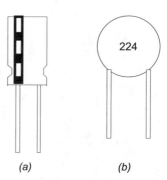

(a)   (b)

Fig. 8 Model of a
capacitor and its parasitics

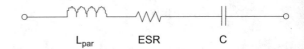

$$L_{par} \qquad ESR \qquad C$$

3 Bypass Capacitors

We have seen that CMOS logic gates, when switching, will draw current from the power supply for a brief amount of time. Since the current is changing rapidly with time, this will induce a voltage in the self-inductance of the wires connected to the power supply. In turn, significant power and ground noise can be generated, to the point of disrupting proper circuit function.

To reduce the power and ground noise, bypass capacitors are used. These capacitors provide a reservoir of charge for each IC; a sudden increase in current consumption can be provided by this reservoir. As a result, the rapid current change is not provided through the power supply wires (inductance), which reduces the noise. Furthermore, for bypass capacitors to be most effective, they must be placed as close as possible to the power pin of the IC (Fig. 9).

In the example shown, the surface mount capacitor is connected directly to the IC's power pin, with a low impedance ground plane serving as the return path. Each IC in the circuit needs its own bypass capacitor, with typical starting values of 0.1 µF.

For the bypass capacitor to be effective, we must be aware of its self-resonance frequency (SRF). A 0.1 µF surface mount capacitor typically has a SRF of 10–20 MHz, making them a good choice for lower-frequency electronics. However, depending on the frequency spectrum of the supply current, other capacitors may need to be used. For example, if the IC's current changes very rapidly, then a 100 pF capacitor with SRF of around 1 GHz may be used to reduce this high-frequency noise.

While a smaller capacitor with a higher SRF can help with rapid current changes, the reduced capacitance does mean the reservoir is smaller, and less charge is available. Hence, a general strategy in electronic systems is to have local bypass capacitors for each IC and then bigger capacitors on the main power bus to supply larger slow-changing currents. This strategy is illustrated in below (Fig. 10).

Fig. 9 Bypass capacitor
placement

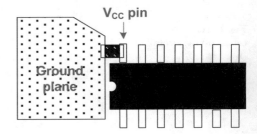

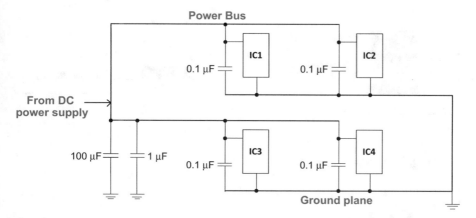

Fig. 10 Bypass capacitor strategy in an electronic system

4   Printed Circuit Boards

Printed circuit boards (PCBs) provide mechanical support for electronic components and the conductors connecting them. At its simplest, a PCB consists of a substrate material such as FR4, on top of which copper tracks and pads are deposited. Electronic components such as ICs are soldered to the copper pads, completing the connection between components. As described, this is a single-layer PCB (Fig. 11).

With only a single layer for routing the copper wires, it may be difficult to complete all the connections. In addition, while bypass capacitors can reduce the effect of power and ground inductances, it is still desirable to minimize this inductance in the first place. One solution is to use a two-layer PCB, where copper is deposited on the top and bottom of the substrate. This not only ease the difficulty of routing connections, but where possible large areas of copper can be connected to ground in the bottom layer to create a *ground plane*. A ground plane provides a low inductance return path for the supply current. To connect traces on the top layer to the bottom layer, holes are drilled into the board and coated with copper to complete the connection; these copper plated holes are known as *vias* (Fig. 12).

The thickness of the copper tracks determines the resistance and the current handling ability of the connection. This is especially important in power electronics, where large currents can flow between components. For a given thickness, making the copper trace wider will increase its maximum current limit. For high currents, however, the wide trace may take up a lot of space on the PCB. Hence, industry standards exist for specifying the copper thickness, in the weight of copper per square foot. 0.5 oz, 1 oz, and 2 oz copper are commonly used, with 1 ounce copper having a thickness of about 34 μm.

As electronic circuits became more complicated and operate at higher frequencies, multilayer PCBs have become more common. For example, in a four-layer PCB, the top and bottom layers can be used for signal tracks, whereas the inner two layers can be used for power and ground planes (Fig. 13).

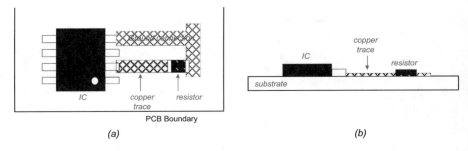

Fig. 11 Single layer PCB (**a**) top view (**b**) side view

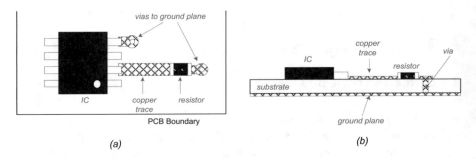

Fig. 12 Two layer PCB (**a**) top view (**b**) side view

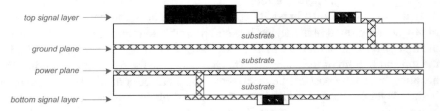

Fig. 13 Four layer PCB, side view

5   Problems

1. For passive components such as resistors and capacitors, what are the two major categories of packages?
2. What is the name of a common thru-hole package for integrated circuits?
3. Compare and contrast the two surface mount packages for integrated circuits: quad flat pack (QFP) versus ball grid array (BGA).
4. What type of capacitor has a polarity associated with it? What happens if you apply the wrong voltage polarity to this type of capacitor?
5. What are the purposes of a bypass capacitor? (checked if already asked in earlier chapter)

6. In terms of physical placement, where do you put the bypass capacitors relative to the ICs in your circuit? What is the reason for this?
7. Why are different capacitor values used for bypass capacitors?
8. In the context of a printed circuit board (PCB), what is a via?
9. When do engineers create ground planes on PCBs?
10. What are the considerations in choosing the copper thickness for a PCB?

Index

© Springer Nature Switzerland AG 2022
C. Siu, *Electronic Devices, Circuits, and Applications*,
https://doi.org/10.1007/978-3-030-80538-8

Printed in the United States
by Baker & Taylor Publisher Services